Dynamical Systems Theory

Edited by Jan Awrejcewicz
and Dariusz Grzelczyk

Published in London, United Kingdom

IntechOpen

Supporting open minds since 2005

Dynamical Systems Theory
http://dx.doi.org/10.5772/intechopen.83328
Edited by Jan Awrejcewicz and Dariusz Grzelczyk

Contributors
Mohamed Abdelsabour Fahmy, Hatem Oueslati, Salah Ben Mabrouk, Abdelkader Mami, Miguel Ramirez, Fadi Dohnal, Joaquin Collado, Josep Sardanyés, Lluís Alsedà, J. Tomás Lázaro, Blai Vidiella, Dariusz Grzelczyk, Jan Awrejcewicz, Lev Steinberg, Roman Kvasov, Olesia Maksymovych, Kesong Yan, Fanping Zeng, Alexandre de Macêdo Wahrhaftig

First published in London, United Kingdom, 2020 by IntechOpen
IntechOpen is the global imprint of INTECHOPEN LIMITED, registered in England and Wales, registration number: 11086078, 7th floor, 10 Lower Thames Street, London, EC3R 6AF, United Kingdom
Printed in Croatia

British Library Cataloguing-in-Publication Data
A catalogue record for this book is available from the British Library

Additional hard and PDF copies can be obtained from orders@intechopen.com

Dynamical Systems Theory
Edited by Jan Awrejcewicz and Dariusz Grzelczyk
p. cm.
Print ISBN 978-1-83880-229-5
Online ISBN 978-1-83880-230-1
eBook (PDF) ISBN 978-1-83880-457-2

For EU product safety concerns:
IN TECH d.o.o., Prolaz Marije Krucifikse Kozulić 3, 51000 Rijeka, Croatia, info@intechopen.com or visit our website at intechopen.com.

Meet the editors

Professor Jan Awrejcewicz is a full professor and Head of the Department of Automation, Biomechanics and Mechatronics at Lodz University of Technology, Poland. He is a doctor honoris causa, recipient of the Humboldt Award, member of the Polish Academy of Sciences, Head of the PhD School on Mechanics, and Head of Graduate/Postgraduate Programs on Mechatronics. He is the author/coauthor of numerous journal papers, conference papers, chapters in books, and 51 monographs. He is the editor of 28 books and 28 journal special issues. He is a member of the editorial boards of 73 journals and principal investigator in 39 research grants. His research interest covers mechanics, biomechanics, automation, physics, and computer-oriented sciences, with a main focus on nonlinear processes. For more see: www.abm.p.lodz.pl.

Dariusz Grzelczyk, PhD, has since 2010 been an employee at the Department of Automation, Biomechanics and Mechatronics at Lodz University of Technology, Poland. He is a member of the Section of Dynamics of Systems of the Polish Academy of Sciences, and a member of the Polish Society of Biomechanics. He is coauthor of many journal papers, conference papers, and chapters in books. He is also guest coeditor of three journal special issues, investigator in eight research grants, and supervisor/cosupervisor of three finalized PhD theses and 17 master/bachelor theses. His research interest covers mechanics, biomechanics, automation, robotics, mechatronics, physics, and computer-oriented sciences. For more see: www.abm.p.lodz.pl.

Contents

Preface XI

Chapter 1 1
Coupled Mathieu Equations: γ-Hamiltonian and μ-Symplectic
by Miguel Ramírez Barrios, Joaquín Collado and Fadi Dohnal

Chapter 2 23
Partial Entropy and Bundle-Like Entropy for Topological Dynamical Systems
by Kesong Yan and Fanping Zeng

Chapter 3 43
Determination of Stresses in Composite Plates with Holes and Cracks Based on Singular Integral Equations
by Olesia Maksymovych and Adam Podhorecki

Chapter 4 63
A New Computerized Boundary Element Model for Three-Temperature Nonlinear Generalized Thermoelastic Stresses in Anisotropic Circular Cylindrical Plate Structures
by Mohamed Abdelsabour Fahmy

Chapter 5 81
Distinctive Characteristics of Cosserat Plate Free Vibrations
by Lev Steinberg and Roman Kvasov

Chapter 6 103
On the Controlling of Multi-Legged Walking Robots on Stable and Unstable Ground
by Dariusz Grzelczyk and Jan Awrejcewicz

Chapter 7 117
On Dynamics and Invariant Sets in Predator-Prey Maps
by Blai Vidiella, J. Tomás Lázaro, Lluís Alsedà and Josep Sardanyés

Chapter 8 139
Using Dynamic Analysis to Adjust the Rheological Model of Three Parameters to the Eurocode Creep Criteria
by Alexandre de Macêdo Wahrhaftig

Chapter 9 155
Dynamic Modelling by Bond Graph Approach of Convective Drying Phenomena
by Hatem Oueslati, Salah Ben Mabrouk and Abdelkader Mami

Preface

The quest to ensure perfect dynamical properties and the control of different systems used in both human everyday life and engineering applications is currently the goal of numerous research conducted by different scientific centers and laboratories all over the world. To help support this effort, various scientific agencies and foundations have established funding for investigations regarding modeling, numerical simulation, and experimental verifications of proposed mathematical models of these systems. As a result of extensive research, numerous interesting solutions allowing for the simulation and analysis of real dynamical systems have been obtained. Different theorems and algorithms have been developed to analyze systems that are usually governed by ordinary or partial differential equations. Their dynamics and stability have been investigated by bifurcation diagrams, Lyapunov exponents, frequency spectra, Fourier transforms, wavelets, asymptotic methods, multiple scale methods, finite element methods, and many other mathematical approaches.

The aim of this book is to provide the reader with a selection of modern methods in the field of mathematical modeling, simulation, dynamics and control of lumped mass mechanical systems, structural members, continuous/discontinuous systems, and parametric physical systems. The book presents selected approaches suitable for the analysis of different kinds of dynamical systems, and is sectioned into nine chapters introducing the reader to different issues of dynamical systems theory. A brief description of each chapter follows.

In Chapter 1, Barrios et al. focus on studies related to the stability transition curves of coupled Mathieu equations utilizing numerical and theoretical methods. They extend previously developed theory and mathematical formalism used for Hamiltonian systems to carry out the investigation of two coupled and damped Mathieu equations used as an example, highlighting advantages of the presented formalism. Stability of the considered system is analyzed by using the appropriate stable/unstable regions where parametric resonance phenomena are highlighted. Due to the symmetry of the used symplectic matrices, the parametric resonance zones are characterized yielding faster and more robust computations, and achieving a higher accuracy while estimating the stability transition curves. The proposed procedure can be extended to a higher number of coupled Mathieu equations.

In Chapter 2, Yan and Zeng consider the problem of the entropy of topological dynamical systems, which is an important notion for understanding the complexity of such systems. The authors study first the relationship among topological entropy, pseudo-orbit, and preimage entropies for topological dynamical systems regarding localization phenomena. Second, they introduce and investigate two entropy-like invariants for non-autonomous discrete dynamical systems, i.e. partial entropy and bundle-like entropy. As a result, the relation between the topological entropy for open covers and several preimage entropy invariants is studied, which is viewed as the local version of the Hurley inequality. Eventually, the authors show that topological entropy for open covers can be computed by measuring the exponential growth rate of the number of pseudo-orbits that end at a particular point.

In Chapter 3, Maksymovych and Podhorecki consider the problems of determination of stresses at cracks in bounded plates with holes of different shapes under the action of concentrated forces or distributed forces at their boundary. The presented approach is developed by using the singular integral equations determined based on the established interdependences between the Lekhnitskii potentials and stress/strain. The numerical method for solving the obtained integral equations is developed based on the quadrature method for the systems exhibiting holes and cracks. The developed numerical algorithm is applied for calculating stresses at cracks in bounded plates with holes of various shapes due to concentrated forces or distributed forces. Eventually, the analysis of stresses at cracks in samples, which are used in experimental studies of crack fracture resistance, is performed.

Fahmy proposes a theory devoted to nonlinear generalized thermoelastic stresses in anisotropic circular cylindrical plate structures in Chapter 4. Due to strong nonlinearity of the considered problem and because of difficulties with finding its analytical solution, a novel computerized boundary element model is developed. To describe the thermal stresses in such structures, two-dimensional temperature nonlinear radiative heat conduction equations coupled with electron, ion, and phonon temperatures are used. To verify the proposed model, the three-temperature radiative heat conduction results are replaced with one-temperature heat conduction results. The presented data obtained by using the developed computerized boundary element model and a finite element model analyzed in COMSOL Multiphysics commercial software confirms the validity and accuracy of the proposed theory.

In Chapter 5, Steinberg and Kvasov present theoretical analysis of the distinctive characteristics of Cosserat plate-free vibrations by using the proposed dynamic model developed as an extension of the Reissner plate theory. During numerical validation of the model the authors detect and classify different characteristics of the plate vibrations, including additional high-resonance frequencies of a plate depending on the shape and orientation of the microelements incorporated into the Cosserat plates. They show that microfrequencies associated with the microrotatory inertia and its transverse variation of the ellipsoid elements have higher microfrequencies than the ball-shaped elements. In addition, the dependence of eigenfrequencies on the angles of rotation of the horizontal ellipsoid microelements is detected.

Chapter 6 looks at Grzelczyk and Awrejcewicz develop and investigate numerically a general kinematic model of a multilegged hybrid robot with crab-like and/or mammal-like legs. The locomotion process and stabilization of the position and orientation of the robot on planar, unstable, and vibrating ground are visualized in Mathematica. The authors focus especially on precise control of the position of the robot during walking in different directions and precisely control all six spatial degrees of freedom of the robot's body, as well as all the robot's legs. The obtained results show the possibility for full control of the robot position and orientation in space, as well as during the locomotion process. The control algorithm can also be successfully employed on unstable or vibrating ground to stabilize the spatial position of the robot.

In Chapter 7, Vidiella et al. analyze the numerical dynamics of a predator-prey biological ecosystem existing in nature and governed by nonlinear difference equations. Despite the relatively simple model of the investigated dynamical system, it is characterized by an enormous richness of dynamics, including extinctions, coextinctions, ordered and chaotic coexistence, as well as hyperchaotic behavior detected by two positive Lyapunov exponents for some parameters' regions. The authors provide conditions for the global stability of the fixed points corresponding to the coextinctions of the predator-prey as well as for the extinction of predators and survival of prey. By iterating the analyzed dynamical system, a very complicated shape of the escaping regions, presumably with a highly entangled fractal topology, is detected and characterized.

Chapter 8 sees Wahrhaftig perform a dynamic analysis of vibration of a real slender reinforced concrete pole with variable geometry, with a concentrated force present at the free end of the structural element. During simulations they consider the self-weight of the structure and nonlinear properties of the material. Using two simulation models, the author demonstrates that the used rheological model characterized by three parameters is suitable for mathematical implementation, in comparison with the standard criteria. It stands for the possibility of adjustment of a simple model to the standard one and shows practical applications to calculate the first natural frequency as the critical buckling load. Eventually, it is noted that in further investigations, a programing routine for obtaining a finer adjustment of the curve between the viscoelastic rheological model and that of the model for creep as predicted by Eurocode criteria should be developed.

Finally, in Chapter 9, Oueslati et al. present a modeling study of heat transfer during the drying of a moist agricultural product placed in a hot air flow in a tunnel dryer with partial solar heating. To do this, they use a bond graph approach, which is an object-oriented graphical approach based on an energetic description between subsystems. Experimental drying tests are carried out to validate the developed model, taking into account the geometry of the dryer, the physical properties of building materials, agricultural product, and air properties. Especially, the influence of temperature and velocity of hot air are studied to evaluate the performance of the dryer. As a result of the optimization technique, the developed model can be adapted to different agricultural products providing their high quality, or to other drying processes.

We hope that the readers of this book will be attracted by the topics covered in the content, which are directed primarily towards research aimed at increasing their academic knowledge with competences related to the selected new mathematical theoretical approaches and original numerical tools related to a few problems of dynamical systems theory.

Jan Awrejcewicz and Dariusz Grzelczyk
Department of Automation, Biomechanics and Mechatronics,
Lodz University of Technology,
Lodz, Poland

Chapter 1

Coupled Mathieu Equations: γ-Hamiltonian and μ-Symplectic

Miguel Ramírez Barrios, Joaquín Collado and Fadi Dohnal

Abstract

Several theoretical studies deal with the stability transition curves of coupled and damped Mathieu equations utilizing numerical and asymptotic methods. In this contribution, we exploit the fact that symplectic maps describe the dynamics of Hamiltonian systems. Starting with a Hamiltonian system, a particular dissipation is introduced, which allows the extension of Hamiltonian and symplectic matrices to more general γ-Hamiltonian and μ-symplectic matrices. A proof is given that the state transition matrix of any γ-Hamiltonian system is μ-symplectic. Combined with Floquet theory, the symmetry of the Floquet multipliers with respect to a μ-circle, which is different from the unit circle, is highlighted. An attempt is made for generalizing the particular dissipation to a more general form. The methodology is applied for calculation of the stability transition curves of an example system of two coupled and damped Mathieu equations.

Keywords: Hamiltonian systems, periodic systems, Mathieu equation, parametric excitation, parametric resonance, symplectic maps

1. Introduction

Dynamical systems represented by nonlinear or linear ordinary differential equations with periodic coefficients occur in many engineer problems (see for instance [1, 2]). The simplest example of such a system is the Mathieu equation. Most investigations in literature deal with the corresponding stability transition curves [3]. Some works analyze the stability of two coupled Mathieu equations [4–6]. In general, an asymptotic or a numerical analysis method is required for analyzing this class of systems. Perturbation techniques may lead to cumbersome expression, at least for second-order perturbation [7], and a numerical analysis may require considerable computation time. In this contribution, an extension of the theory developed in [8] is exposed in which coupled Mathieu equations are analyzed in the context of a Hamiltonian system.

The literature on Hamiltonian systems is vast. We focus on the two main references [9, 10] that are relevant for the present work. The latter focuses on linear periodic Hamiltonian systems. Although every periodic mechanical system possesses at least a small amount of dissipation, the main literature on linear Hamiltonian systems does not incorporate a dissipation. The dynamics of Hamiltonian systems can be described by symplectic maps [11]. A key fact here is that a symplectic transformation preserves the Hamiltonian structure of the underlying dynamic system. In this work we attempt to derive an appropriate formalism for linear Hamiltonian systems incorporating a very particular dissipation. For this purpose we redefine and develop the properties of the so-called γ-Hamiltonian and μ-symplectic matrices. With the last definitions, we prove that the state transition matrix of any γ-Hamiltonian system is μ-symplectic. The relevance of the symplectic matrices or symplectic maps lies on their symmetry which allows simplifying many computations and analysis [12]. The formalism is benchmarked for two coupled and damped Mathieu equations highlighting its advantages. Due to the symmetry of the symplectic matrices, the parametric resonance zones are characterized, which allows faster computations, and with higher accuracy, of the stability transition curves. This work is an extension of the contribution presented in [8, 13].

2. Preliminaries on matrices

2.1 Symplectic matrices

Definition 1 *The matrix $A \in \mathbb{R}^{2n\times 2n}$ is called symplectic if it satisfies*

$$A^T J A = J, \tag{1}$$

with

$$J = \begin{bmatrix} 0 & I_n \\ -I_n & 0 \end{bmatrix} \tag{2}$$

and I_n is the $n \times n$ identity matrix.

Note that for J the following relations hold: $J^T = -J, J^{-1} = J^T, J^2 = -I_{2n}$, and $\det(J) = 1$. The determinant of a symplectic matrix is 1 ([9]), and I_{2n} and J are symplectic matrices themselves. If A and B are of the same dimensions and symplectic, then AB is also symplectic because $(AB)^T J(AB) = B^T A^T JAB = B^T JB = J$. Finally and importantly, the inverse of a symplectic matrix always exists and is also symplectic:

$$A^{-1} = J^{-1}A^T J : \qquad (J^{-1}A^T J)^T J (J^{-1}A^T J) = J^T AJA^T J = J. \tag{3}$$

The set of the symplectic matrices of dimension $2n \times 2n$ forms a group. The corresponding characteristic polynomial of a symplectic matrix $A \in \mathbb{R}^{2n\times 2n}$

$$P_A(\lambda) = \det(\lambda I_{2n} - A) = \lambda^{2n} + a_{2n-1}\lambda^{2n-1} + \ldots + a_1\lambda + 1$$

is a reciprocal polynomial:

$$P_A(\lambda) = \lambda^{2n} P_A\left(\frac{1}{\lambda}\right) \tag{4}$$

This is equivalent to stating that the coefficients of $P_A(\lambda)$ satisfy the relation $a_k = a_{2n-k}$ or rewriting as a matrix product

$$\begin{bmatrix} a_0 \\ a_1 \\ \vdots \\ a_{2n-1} \\ a_{2n} \end{bmatrix} = \begin{bmatrix} 0 & 0 & \cdots & 0 & \cdots & 0 & 1 \\ 0 & 0 & \cdots & 0 & \cdots & 1 & 0 \\ \vdots & \vdots & \ddots & \vdots & \ddots & \vdots & \vdots \\ 0 & 0 & \cdots & 1 & \cdots & 0 & 0 \\ \vdots & \vdots & \ddots & \vdots & \ddots & \vdots & \vdots \\ 0 & 1 & \cdots & 0 & \cdots & 0 & 0 \\ 1 & 0 & \cdots & 0 & \cdots & 0 & 0 \end{bmatrix} \begin{bmatrix} a_0 \\ a_1 \\ \vdots \\ a_{2n-1} \\ a_{2n} \end{bmatrix} \tag{5}$$

Since A is real, if λ is an eigenvalue of A, then so are λ^{-1}, $\bar{\lambda}$, and $\bar{\lambda}^{-1}$, where the bar indicates the complex conjugate. Equivalently, the eigenvalues of a symplectic matrix are reciprocal pairs. This property is called reflexivity [11]. Consequently, the eigenvalues are symmetric with respect to the unit circle, namely, if there is an eigenvalue inside of the unit circle, then there must be a corresponding eigenvalue outside of the unit circle. As a result of the coefficient symmetry of a symplectic matrix A, the following transformation is proposed in [12]:

$$\delta = \lambda + \frac{1}{\lambda}, \tag{6}$$

where $\lambda \in \sigma(A)$. This transforms the characteristic polynomial $P_A(\lambda)$ of degree $2n$ to an auxiliary polynomial $Q_A(\delta)$ of degree n, while keeping all pertinent information of the original polynomial [12].

2.2 Hamiltonian matrices

Definition 2 *The matrix* $A \in \mathbb{R}^{2n\times 2n}$ $(A \in \mathbb{C}^{2n\times 2n})$ *is said to be Hamiltonian if and only if*

$$A^T J + JA = \mathbf{0}. \tag{7}$$

Let $P_A(s)$ be the characteristic polynomial of A, then $P_A(s)$ is an even polynomial, and it only has even powers. Thus, the eigenvalues of A are symmetric with respect to the imaginary axis, i.e., if s is an eigenvalue of A, then $-s$ is an eigenvalue, too. Furthermore, if the matrix A is real, $\bar{s}$ and $-\bar{s}$ are eigenvalues as well. Then the eigenvalues of the Hamiltonian matrix are located symmetrically with respect to both real and imaginary axis. The eigenvalues appear in real pairs, purely imaginary pairs, or complex quadruples [9, 14].

2.3 μ-symplectic matrices

The next definitions and properties attempt to generalize the classical definitions above.

Definition 3 $M \in \mathbb{R}^{2n\times 2n}$ *is called* μ*-symplectic matrix if*

$$M^T J M = \mu J \tag{8}$$

is satisfied for $\mu \in (0, 1]$.

Lemma 4 *The determinant of a* μ*-symplectic matrix* $M \in \mathbb{R}^{2n\times 2n}$ *is* μ^n.

To see the proof of the last lemma, see Appendix A. If M is a μ-symplectic matrix, M^2 is a μ^2-symplectic matrix, and the set of μ-symplectic matrix matrices does not form a group.

Lemma 5 *The characteristic polynomial of a μ-symplectic matrix $M \in \mathbb{R}^{2n\times 2n}$ satisfies*

$$P_M\left(\frac{\mu}{\lambda}\right) = \frac{\mu^n}{\lambda^{2n}} P(\lambda). \tag{9}$$

Proof 6 $P_M(\lambda) = \det(\lambda I_{2n} - M^T) = \det(\lambda I_{2n} - \mu J M^{-1} J^{-1})$

$$= \det(J)\det(\lambda I_{2n} - \mu M^{-1})\det(J^{-1}) = \det\left(\frac{\lambda}{\mu}M - I_{2n}\right)\det(\mu M^{-1})$$

$$= \mu^n \det\left(\left(-\frac{\lambda}{\mu}\right)\left(\frac{\mu}{\lambda}I_{2n} - M\right)\right) = \mu^n\left(-\frac{\lambda}{\mu}\right)^{2n}\det\left(\frac{\lambda}{\mu}I_{2n} - M\right) = \frac{\lambda^{2n}}{\mu^n}P_M\left(\frac{\mu}{\lambda}\right)$$

Corollary 7 *The eigenvalues of a μ-symplectic matrix M satisfy the symmetry*

$$\lambda \in \sigma(M) \Rightarrow \left(\frac{\mu}{\lambda}\right) \in \sigma(M). \tag{10}$$

The product of each pair of eigenvalues contributes with μ to $\det(M)$, *and there are n of these pairs; therefore,* $\det(M) = \mu^n$. *If all eigenvalues have the same magnitude, i.e., $\lambda_i = r\exp(\theta_i)$, then $\prod_{i=1}^{2n}|\lambda| = \prod_{i=1}^{2n}\left|re^{\theta_i}\right| = r^{2n} = \det(M) = \mu^n$. From this we find that $r = \sqrt{\mu}$, independent of n. This may be interpreted as a "symmetry" with respect to a circle of radius $r = \sqrt{\mu}$. Since M is real if λ is an eigenvalue of M, then $\overline{\lambda}$, $\frac{\mu}{\lambda}$, and $\frac{\mu}{\overline{\lambda}}$ are also eigenvalues of M. Moreover, the eigenvalues are symmetric with respect to the μ-circle: if there is an eigenvalue inside of the μ-circle, then there must be another eigenvalue outside (see* **Figure 1a** *for a visualization).*

Remark 8 *Due to Eq. (9), the characteristic polynomial $P_M(\lambda) = m_{2n}\lambda^{2n} + \ldots m_1\lambda + m_0$ of the μ-symplectic matrix M satisfies the following relations:*

$$\begin{aligned} m_0 &= m_{2n}\mu^n \\ m_1 &= m_{2n-1}\mu^{n-1} \\ &\vdots \\ m_n &= m_n \\ &\vdots \\ m_{2n-1} &= m_1\mu^{1-n} \\ m_{2n} &= 1 = m_0\mu^{-n} \end{aligned} \tag{11}$$

rewritten as a product of matrices yields

$$\begin{bmatrix} m_0 \\ m_1 \\ \vdots \\ m_{2n-1} \\ m_{2n} \end{bmatrix} = \begin{bmatrix} 0 & 0 & \cdots & 0 & \cdots & 0 & \mu^{-n} \\ 0 & 0 & \cdots & 0 & \cdots & \mu^{-n+1} & 0 \\ \vdots & \vdots & \ddots & \vdots & \ddots & \vdots & \vdots \\ 0 & 0 & \cdots & 1 & \cdots & 0 & 0 \\ \vdots & \vdots & \ddots & \vdots & \ddots & \vdots & \vdots \\ 0 & \mu^{n-1} & \cdots & 0 & \cdots & 0 & 0 \\ \mu^n & 0 & \cdots & 0 & \cdots & 0 & 0 \end{bmatrix} \begin{bmatrix} m_0 \\ m_1 \\ \vdots \\ m_{2n-1} \\ m_{2n} \end{bmatrix} \tag{12}$$

For $\mu = 1$, the relations in Eq. (12) reduce to Eq. (5).

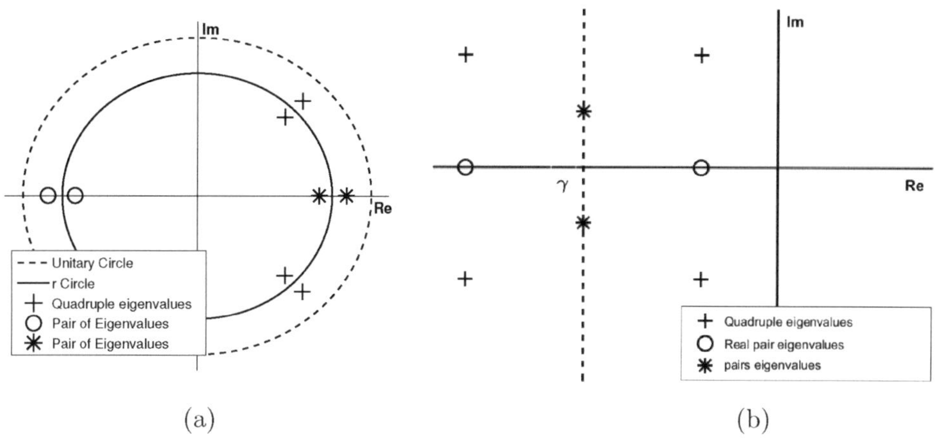

Figure 1.
Symmetries in the spectra in the complex plane: (a) μ-symplectic matrix and (b) γ-Hamiltonian matrix.

Remark 9 *By applying the transformation*

$$\delta = \lambda + \frac{\mu}{\lambda} \tag{13}$$

the characteristic polynomial $P_M(\lambda)$ of degree $2n$, associated to a μ-symplectic matrix, is reduced to an auxiliary polynomial $Q_M(\delta)$ of degree n. For instance,

$$n = 2 \quad \begin{cases} P_M(\lambda) = \lambda^4 + m_3\lambda^3 + m_2\lambda^2 + m_3\mu\lambda + \mu^2 \\ Q_M(\delta) = \delta^2 + m_3\delta + m_2 - 2\mu \end{cases}$$

$$n = 3 \quad \begin{cases} P_M(\lambda) = \lambda^6 + m_5\lambda^5 + m_4\lambda^4 + m_3\lambda^3 + \mu m_4\lambda^2 + \mu^2 m_5\lambda + \mu^3 \\ Q_M(\delta) = \delta^3 + m_5^2\delta + (m_4 - 3\mu)\delta + m_3 - 2m_5\mu \end{cases}$$

$$n = 4 \quad \begin{cases} P_M(\lambda) = \lambda^8 + m_7\lambda^7 + m_6\lambda^6 + m_5\lambda^5 + m_4\lambda^4 + \mu m_5\lambda^3 \\ \qquad + \mu^2 m_6\lambda^2 + \mu^3 m_7\lambda + \mu^4 \\ Q_M(\delta) = \delta^4 + m_7^3\delta + (m_6 - 4\mu)\delta^2 + (m_5 - 3m_7\mu)\delta + m_4 \\ \qquad - 2m_6\mu + 2\mu^2 \end{cases}$$

$$n = 5 \quad \begin{cases} P_M(\lambda) = \lambda^{10} + m_9\lambda^9 + m_8\lambda^8 + m_7\lambda^7 + m_6\lambda^6 + m_5\lambda^5 \\ \qquad + \mu m_6\lambda^4 + \mu^2 m_7\lambda^3 + \mu^3 m_8\lambda^2 + \mu^4 m_9\lambda + \mu^5 \\ Q_M(\delta) = \delta^5 + m_9\delta^4 + (m_8 - 5\mu)\delta^3 + (m_7 - 4m_9\mu)\delta^2 \\ \qquad + (m_6 - 3m_8\mu + 5\mu^2)\delta + m_5 - 2\mu m_7 + 2m_9\mu^2 \end{cases}$$

Note that the property of the characteristic polynomial of a μ-symplectic matrix in Eq. (9) reduces to Eq. (4) at $\mu = 1$. Then Eq. (12) represents the "symmetry" of the characteristic polynomial for all $\mu \in (0, 1]$. Although the definition of μ-symplectic matrices appears in [9], no further properties were developed within this reference. In the next section, we reveal their relationship as a generalized definition of Hamiltonian matrices, the so-called γ-Hamiltonian matrices.

2.4 γ-Hamiltonian matrices

Definition 10 *A matrix* $A \in \mathbb{R}^{2n\times 2n}$ ($A \in \mathbb{C}^{2n\times 2n}$) *is called* γ*-Hamiltonian matrix if for some* $\gamma \geq 0$,

$$A^T J + JA = -2\gamma J \tag{14}$$

Lemma 11 *A is* γ*-Hamiltonian if and only if* $A + \gamma I_{2n}$ *is Hamiltonian.*

Proof 12 *If A is* γ*-Hamiltonian, then* $A^T J + JA = -2\gamma J$ *which can be rewritten as* $[A + \gamma I_{2n}]^T J + J[A + \gamma I_{2n}] = 0$. *Hence,* $[A + \gamma I_{2n}]$ *is Hamiltonian.*

Lemma 13 *If A is* γ*-Hamiltonian and if* $s + \gamma \in \sigma(A)$, *then* $-s + \gamma \in \sigma(A)$.

Proof 14 *Recall that if* $\sigma(R) = \{r_1, ...r_{2n}\}$, *then* $\sigma(R + \gamma I_{2n}) = \{r_1 + \gamma, ..., r_{2n} + \gamma\}$. *Then if* $s + \gamma \in \sigma(A)$, *then* $s \in \sigma(A + \gamma I_{2n})$, *since* $[A + \gamma I_{2n}]$ *is Hamiltonian and consequently* $-s \in \sigma(A + \gamma I_{2n})$ *which is equivalent to* $-s + \gamma \in \sigma(A)$.

Remark 15 *If in the last lemma all the eigenvalues of the Hamiltonian matrix* $A + \gamma I_{2n}$ *have zero real parts, then the real parts of the eigenvalues of the* γ*-Hamiltonian matrix A are identical to* $-\gamma$. *Thus, the eigenvalues of the* γ*-Hamiltonian matrix A are symmetric with respect to the vertical line* $-\gamma$ *in the complex plane (see* **Figure 1b** *for a visualization).*

Notice that real Hamiltonian matrices have their spectrum symmetric with respect to the real and imaginary axes, whereas the spectrum of real γ-Hamiltonian matrices is symmetric with respect to the real axis and a vertical line at $\mathbf{Re}(s) = -\gamma$. Then the eigenvalues of a real γ-Hamiltonian matrix are placed: (*i*) in quadruples symmetrically with respect the real axis and the line $\mathbf{Re}(s) = -\gamma$, (*ii*) pairs on the line $\mathbf{Re}(s) = -\gamma$ and symmetric with the real axis, and (*iii*) real pairs symmetric with the line $\mathbf{Re}(s) = -\gamma$. All cases are shown in **Figure 1b**.

By the last lemma, the characteristic polynomial of the γ-Hamiltonian A satisfies

$$P_A(s + \gamma) = P_A(\gamma - s)$$

with

$$P_A(\gamma - s) = (\gamma - s)^{2n} + a_{2n-1}(\gamma - s)^{2n-1} + ... + a_1(\gamma - s) + a_0$$
$$P_A(s + \gamma) = (s + \gamma)^{2n} + a_{2n-1}(s + \gamma)^{2n-1} + ... + a_1(s + \gamma) + a_0$$

Thus, $P_A(s)$ depends only on n coefficients. For instance, for $n = 1$, $(s + \gamma)^2 + a_1(s + \gamma) + a_0 = (\gamma - s)^2 + a_1(\gamma - s) + a_0$. Equating the coefficients leads to $a_1 = -2\gamma$, $a_0 = a_0$, and finally to

$$P_A(s) = s^2 - 2\gamma s + a_0.$$

Similarly, the polynomials for the lowest values of n read

$n = 2:$

$$P_A(s) = s^4 - 4\gamma s^3 + a_2 s^2 + (8\gamma^3 - 2\gamma a_2) + a_0$$

$n = 3:$

$$P_A(s) = s^6 - 6\gamma s^5 + a_4 s^4 + (40\gamma^3 - 4\gamma a_4)s^3 + a_2 s^2 + (-96\gamma^5 + 8\gamma^3 a_4 - 2\gamma a_2) + a_0$$

$n = 4:$

$$P_A(s) = s^8 - 8\gamma s^7 + a_6 s^6 + (112\gamma^3 - 6\gamma a_6)s^5 + a_4 s^4 + (-896\gamma^5 + 40\gamma^3 a_6 - 4\gamma a_4)s^3 + a_2 s^2 + (2176\gamma^7 - 96\gamma^5 a_6 + 8\gamma^3 a_4 - 2\gamma a_2) + a_0.$$

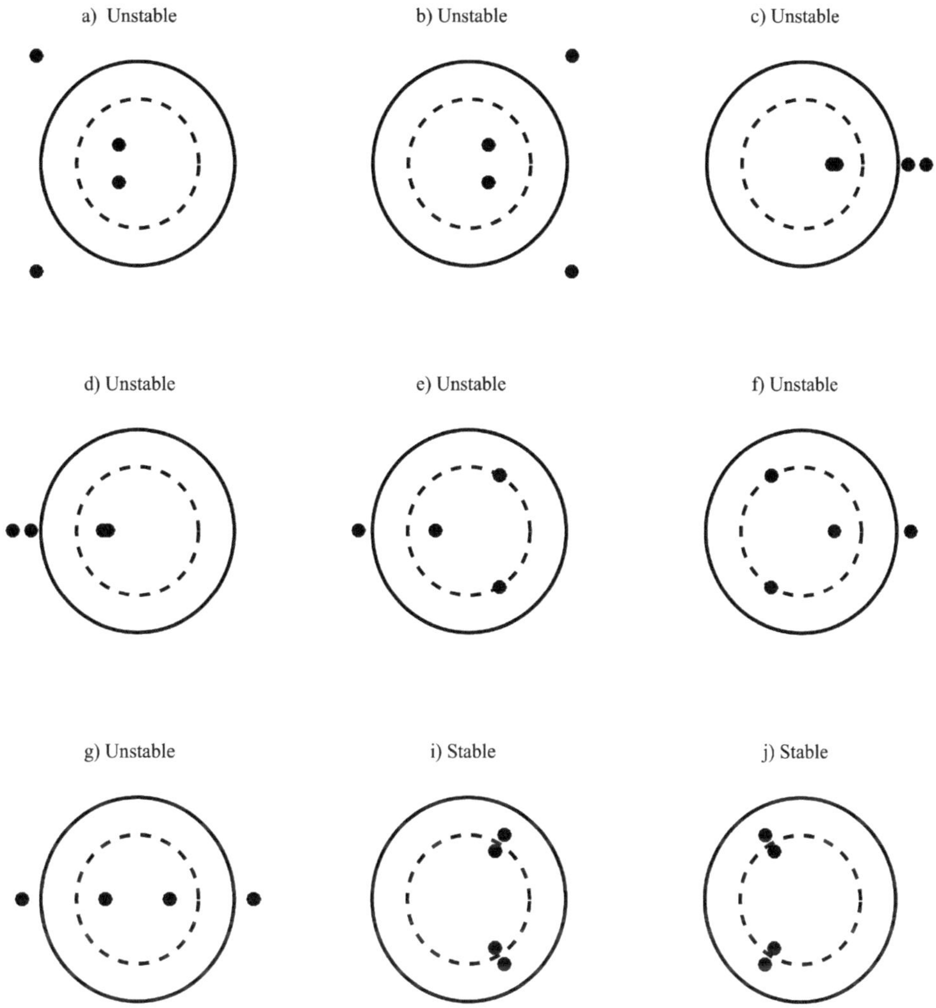

Figure 2.
All configurations of multiplier positions with respect to the unit (in solid line) and μ-circle (dashed line).

Furthermore, by applying the transformation

$$\phi = s - \gamma, \tag{15}$$

the polynomial $P_A(s)$ can be written as an auxiliary polynomial $Q_A(\phi)$ which only has even coefficients, namely,

$$P_A(s) = s^{2n} + a_{2n-1}s^{2n-1} + a_{2n-2}s^{2n-2} + \dots + a_2s^2 + a_1s + a_0$$
$$Q_A(\phi) = \phi^{2n} + q_{2n-2}\phi^{2n-2} + \dots + q_2\phi^2 + q_0$$

For instance,

$$n = 1:$$
$$Q_A(\phi) = \phi^2 + a_0 - \gamma^2$$
$$n = 2:$$
$$Q_A(\phi) = \phi^4 + \left(a_2 - 6\gamma^2\right)\phi^2 + 5\gamma^4 - a_2\gamma^2 + a_0$$

$$n = 3:$$
$$Q_A(\phi) = \phi^6 + \left(a_4 - 15\gamma^2\right)\phi^4 + \left(a_2 - 6a_4\gamma^2 + 75\gamma^4\right)\phi^2 - 61\gamma^6 + 5a_4\gamma^4 - a_2\gamma^2 + a_0$$
$$n = 4:$$
$$Q_A(\phi) = \phi^8 + \left(a_6 - 28\gamma^2\right)\phi^6 + \left(a_4 - 15a_6\gamma^2 + 350\gamma^4\right)\phi^4$$
$$+\left(a_2 - 6a_4\gamma^2 + 75a_6\gamma^4 - 1708\gamma^6\right)\phi^2 + 1385\gamma^8 - 61a_6\gamma^6 + 5a_4\gamma^4 - a_2\gamma^2 + a_0.$$

3. Linear γ-Hamiltonian systems

Definition 16 *If there is a differentiable function called Hamiltonian function (energy) $\mathcal{H}(t,x,y)$, $\mathcal{H}:\mathbb{R}\times\mathbb{R}^n\times\mathbb{R}^n\mapsto\mathbb{R}$, which satisfies*

$$\dot{x} = \left(\frac{\partial\mathcal{H}}{\partial y}\right)^T \text{ and } \dot{y} = -\left(\frac{\partial\mathcal{H}}{\partial y}\right),$$

then it is called a Hamiltonian system. If $\mathcal{H}(t,x,y)$ is a quadratic function with respect to x and y, then the system is a linear Hamiltonian system.

It is easy to prove that if $\mathcal{H}$ does not depend on t, $\mathcal{H}(x,y)$ is a first integral. However, this is no longer true in the time-periodic case. In the time-periodic case, even for $n = 1$, the integration of the equations is not possible. Any linear Hamiltonian system can be written as

$$\dot{z} = JH(t)z \tag{16}$$

where $H^T(t) = H(t)$ is a symmetric matrix (Hermitian in the complex case). Herein, the variables used in the definition satisfy $z = \left[x^T, y^T\right]^T$. Therefore, the dimension of real Hamiltonian systems is always even. Finally, note that the product JH satisfies the condition for a Hamiltonian matrix. The fundamental property of any linear Hamiltonian system is that the state transition matrix of the system in Eq. (16) is a symplectic matrix (see [9] for more details).

If A is γ-Hamiltonian matrix, or equivalently, $A + \gamma I_{2n}$ is a Hamiltonian matrix for some $\gamma > 0$; then it follows from Eq. (16) that

$$\dot{x} = [A + I_{2n}]x = JHx$$

for some matrix $H = H^T$. From the last equation $[A + I_{2n}] = JH$, we obtain

$$A = J[H + \gamma I_{2n}]. \tag{17}$$

Any γ-Hamiltonian matrix A may be written as in Eq. (17), which motivates the next definition.

Definition 17 *Any linear system that can be written as*

$$\dot{x} = A(t)x = J[H(t) + \gamma J]x \tag{18}$$

with $x \in \mathbb{R}^{2n}$, $H^T(t) = H(t)$, and $\gamma \geq 0$ is called a **linear γ-Hamiltonian system**.

Lemma 18 *The state transition matrix of a linear γ-Hamiltonian system in Eq. (18) is μ-symplectic with $\mu = e^{-2\gamma t}$.*

Proof 19 *Let be $N(t) = \Phi(t,0)$ be the state transition of Eq. (17), and then*

$$\dot{N}(t) = A(t)N(t).$$

Differentiating the product N^TJN gives

$$\begin{aligned}\frac{d}{dt}N^TJN &= \dot{N}^TJN + N^TJ\dot{N} = (AN)^TJN + N^TJ(AN) \\ &= N^T(A^TJ + JA)N = N^T\Big((J(H+\gamma J))^TJ + J(J(H+\gamma J))\Big)N \\ &= -2\gamma N^TJN\end{aligned} \tag{19}$$

Since $N^T(0)JN(0) = J$, we get[1]

$$N^T(t)JN(t) = e^{-2\gamma t}J = \mu J.$$

Therefore, N is μ-symplectic.

Lemma 20 *Consider the transformation*

$$x = S(t)z \tag{20}$$

with $S(t)$ a symplectic matrix for all t. Then the transformation in Eq. (20) preserves the γ-Hamiltonian form of the system, Eq.(18).

Proof 21 *From the definition $S^TJS = 0 \rightarrow \dot{S}^TJS + S^TJ\dot{S} = 0$, thus $\dot{S}^TJS = -S^TJ\dot{S}$, and from Eq.(20)*

$$\dot{x} = S\dot{z} + \dot{S}z \rightarrow S^{-1}\dot{x} = \dot{z} + S^{-1}\dot{S}z$$

then applying the transformation Eq.(20) into Eq.(18) it is obtained as $\dot{z} + S^{-1}\dot{S}z = S^{-1}J(H+\gamma J)Sz$; then from the symplectic definition matrix $S^{-1} = J^{-1}S^TJ$,

$$\begin{aligned}\dot{z} &= S^{-1}J(H+\gamma J)Sz - S^{-1}\dot{S}z = (J^{-1}S^TJ)JHSz - \gamma Iz - (J^{-1}S^TJ)\dot{S}z \\ &= JS^THSz - \gamma Iz + JS^TJ\dot{S}z = J(S^THS + S^TJ\dot{S} + \gamma J)z = J(\tilde{H} + \gamma J)z\end{aligned}$$

where $\tilde{H} = S^THS + S^TJ\dot{S}$, but $(S^TJ\dot{S})^T = \dot{S}^TJ^TS = -\dot{S}^TJS = -(-S^TJ\dot{S}) = S^TJ\dot{S}$; therefore $\tilde{H} = \tilde{H}^T$ ■.

3.1 Mechanical, linear γ-Hamiltonian system

Consider any mechanical system described by the equation

$$\tilde{M}\ddot{y} + \tilde{D}\dot{y} + \tilde{K}(t)y = 0 \tag{21}$$

where $y(t) \in \mathbb{R}^n$, $\tilde{K}(t) = \tilde{K}^T(t) \in \mathbb{R}^{n\times n}$, and the constant matrices $\tilde{M}$ and $\tilde{D} \in \mathbb{R}^{n\times n}$ such that $\tilde{M} = \tilde{M}^T > 0$ and $\tilde{D} = \tilde{D}^T$. Then there always exists a linear transformation T such that

$$\begin{aligned}T^T\tilde{M}T &= I_n \\ T^T\tilde{D}T &= D = \text{diag}\{d_1, d_2, \dots, d_n\} \\ \sigma(\tilde{M}^{-1}\tilde{D}) &= \{d_1, d_2, \dots, d_n\}\end{aligned}$$

[1] The matrix product $(\frac{d}{dt}N^TJN)N^TJN = N^TJN(\frac{d}{dt}N^TJN)$ is commutative.

(e.g., see [15]). Therefore, applying the transformation $y = Tz$ yields

$$\ddot{z} + D\dot{z} + K(t)z = 0, \tag{22}$$

where $K(t) = T^T\tilde{K}(t)T$. Eq. (22) can be rewritten as a first-order system by introducing the state vector $x = \left[z^T, \dot{z}^T\right]^T$:

$$\dot{x} = \begin{bmatrix} 0_{n\times n} & I_n \\ -K(t) & -D \end{bmatrix} x \tag{23}$$

where $x \in \mathbb{R}^{2n\times 2n}$. Let

$$Q = \frac{1}{\sqrt{2}}\begin{bmatrix} I_n & I_n \\ -I_n & I_n \end{bmatrix} \tag{24}$$

be an orthogonal matrix satisfying $QQ^T = Q^TQ = I_{2n}$, and also $JQ = Q\,J$, one can introduce the transformation $w = Q^Tx$, and Eq. (23) gives

$$\dot{w} = Q^T\begin{bmatrix} 0_{n\times n} & I_n \\ -K(t) & -D \end{bmatrix} Qw = \frac{1}{2}\begin{bmatrix} K(t) - I_n - D & K(t) + I_n + D \\ -K(t) + D - I_n & -K(t) + I_n - D \end{bmatrix} w,$$

or equivalently,

$$\dot{w} = J\left(\frac{1}{2}\begin{bmatrix} K(t) + I_n - D & K(t) - I_n \\ K(t) - I_n & K(t) + I_n + D \end{bmatrix} + \frac{1}{2}\begin{bmatrix} D & 0_{n\times n} \\ 0_{n\times n} & D \end{bmatrix} J\right) w. \tag{25}$$

Since $D = \operatorname{diag}\{d_1, d_2, \ldots, d_n\}$ and $K = K^T$, the matrix

$$H(t) = \frac{1}{2}\begin{bmatrix} K(t) + I_n - D & K(t) - I_n \\ K(t) - I_n & K(t) + I_n + D \end{bmatrix}$$

is also symmetric $H(t) = H(t)^T$. Therefore, Eq. (25) can be cast into the γ-Hamiltonian linear system form $\dot{w} = J(H + \gamma J)w$ if γ is approximated as $\gamma \approx \frac{1}{2n}\sum_{i=1}^{n} d_i$. In the special case $d = d_1 = d_2 = \cdots = d_n$, γ is given exactly given by $\gamma = \frac{d}{2}$.

3.2 Periodic linear systems

This section summarizes the main results on periodic linear systems. The proofs and details are omitted and can be found in [16, 17]. Consider the linear periodic system:

$$\dot{x} = B(t)x \quad \text{with} \quad B(t) = B(t + \Omega) \tag{26}$$

where $x \in \mathbb{R}^n$, $B \in \mathbb{R}^{n\times n}$, and Ω are the fundamental periods.

Theorem 22 (Floquet) *The state transition matrix $\Phi(t, t_0)$ of the system in Eq. (26) may be factorized as*

$$\Phi(t, t_0) = P^{-1}(t)e^{R(t-t_0)}P(t_0) \tag{27}$$

where

$$P^{-1}(t) = \Phi(t, 0)e^{-Rt}. \tag{28}$$

In addition $P^{-1}(t) = P^{-1}(t+\Omega)$ *is a periodic matrix of the same period* Ω*, and* R *is in general a complex constant matrix [18].*

Definition 23 *We define the monodromy matrix M associated to the Eq. (26) as*

$$M = \Phi(\Omega, 0). \tag{29}$$

The monodromy matrix may be defined as $M_{t_0} = \Phi(\Omega, \Omega + t_0)$, but we use only the spectrum of the monodromy matrix, $\sigma(M)$. From.

$\Phi(t, t_0) = P^{-1}(t)e^{R(t-t_0)}P(t_0)\big|_{t=t_0+\Omega} = \Phi(\Omega, \Omega + t_0) = P^{-1}(t_0 + \Omega)e^{RW}P(t_0)$ $= P^{-1}(t_0)e^{RW}P(t_0)$, because P and P^{-1} are Ω-periodic. This last relation shows that M and M_{t_0} are similar matrices and possess the same spectrum. Moreover, if $t_0 = 0$ in the Floquet theorem, then $\Phi(t, 0) = Q(t)e^{Rt}$ based on $Q(t) = Q(t+\Omega)$ and $Q(0) = I_n$; we have

$$M = \Phi(\Omega, 0) = Q(\Omega)e^{R\Omega} = Q(0)e^{R\Omega} = e^{R\Omega}. \tag{30}$$

Definition 24 *The eigenvalues* λ_i *of the monodromy matrix are called* ***characteristic multipliers*** *or* ***multipliers****. The numbers* ρ_i*, not unique, defined as* $\lambda_i = e^{\rho_i \Omega}$*, are called* ***characteristic exponents*** *or* ***Floquet exponents****.*

Corollary 25 (Lyapunov-Floquet Transformation) *If we define the change of coordinates*

$$z(t) = P(t)x(t) \tag{31}$$

where P fulfills Eq. (28), then the periodic linear system in Eq. (26) can be transformed into a linear time-invariant system

$$\dot{z}(t) = Rz(t) \tag{32}$$

where R is a constant matrix as introduced in the Floquet theorem.

The transformation in Eq. (31) is a **Lyapunov transformation** which means that the stability properties of the linear system in Eq. (26) are preserved. Therefore any periodic system as in Eq. (26) is reducible to a system in Eq. (32) with constant coefficients[2] ([16]). However, the matrix R is not always real (e.g., see [10, 20]). In the present discussion, we only use its spectrum $\sigma(R)$.

For analyzing $x(t)$ as $t \to \infty$, we assume that the initial conditions are given at $t_0 = 0$. Then for any $t > 0$, t may be expressed as $t = k\Omega + \tau$, where $k \in \mathbb{Z}_+$ and $\tau \in [0, \Omega)$. Applying the well-known properties of the state transition matrix, the solution can be written as

$$\begin{aligned} x(t) &= \Phi(t,0)x_0 = \Phi(k\Omega + \tau, 0)x_0 \\ &= \Phi(k\Omega + \tau, k\Omega)\underbrace{\Phi(k\Omega, (k-1)\Omega)\Phi((k-1)\Omega, (k-2)\Omega)\ldots\Phi(\Omega, 0)}_{k-\text{terms}}x_0 \\ &= \Phi(\tau, 0)[\Phi(\Omega, 0)]^k x_0 = \Phi(\tau, 0)M^k x_0 \end{aligned}$$

Analyzing the last expression, the terms $\Phi(\tau, 0)$ and x_0 are bounded; the following three cases can be distinguished:

---

[2] For applying the transformation in Eq. (31), the analytical solution of Eq. (26) is only available for special cases [19], and in general a numerical solution needs to be calculated.

$$x(t) \to 0 \Leftrightarrow \lim_{k\to\infty} M^k = 0 \Leftrightarrow \sigma(M) \subset D = \{z \in \mathbb{C} : |z| < 1\}.$$

1. the solution $x(t)$ is bounded $\Leftrightarrow \lim_{k\to\infty} M^k = 0$ is bounded $\Leftrightarrow \sigma(M) \subset \overline{D} = \{z \in \mathbb{C} : |z| \le 1\}$, and if $\lambda \in \sigma(M)$ and $|\lambda| = 1$, λ is a simple root of the minimal polynomial of M.

2. $\|x(t)\| \to \infty \Leftrightarrow \exists\, \lambda \in \sigma(M) : |\lambda| > 1$ or $\exists\, \lambda \in \sigma(M) : |\lambda| = 1$ and λ is a multiple root of the minimum polynomial of M.

Theorem 26 (Lyapunov-Floquet) *Considering the linear periodic system in Eq. (26), then the system is (a) asymptotic stable if and only if Eq. (1) is satisfied, (b) stable if and only if Eq. (2) is satisfied, and (c) Unstable if and only if Eq. (3) is satisfied.*

Due to the Lyapunov-Floquet transformation in Eq. (31), the stability of the periodic linear system in Eq. (26) can be determined by analyzing the system in Eq. (32).

Corollary 27 *The system in Eq. (26) is:*

i. *Asymptotically stable* $\Leftrightarrow \sigma(R) \subset Z = \{z \in \mathbb{C} : \mathbf{Re}(z) < 0\}$.

ii. *Stable* $\Leftrightarrow \sigma(R) \subset \overline{Z} = \{z \in \mathbb{C} : \mathrm{Re}(z) \le 0\}$, *if* $\mathbf{Re}(z_i) = 0$ *are simple roots of the minimum polynomial of R.*

iii. *Unstable* $\Leftrightarrow \exists\, \rho_i \in \sigma(R) : \mathrm{Re}(z) > 0$ *or* $\sigma(R) \subset \overline{Z}$ & $\exists\, \mathbf{Re}(z_i) = 0$ *which is a multiple root of minimum polynomial of R.*

4. Periodic γ-Hamiltonian systems

Once the linear Hamiltonian systems become periodic, i.e., the matrix $H(t)$ of the system in Eq. (18) possesses a periodically time-varying $H(t) = H(t + \Omega)$, the underlying monodromy matrix becomes μ-symplectic and γ-Hamiltonian.

Definition 28 *Any linear periodic system that can be written as*

$$\dot{x} = J[H(t) + \gamma J]x \tag{33}$$

with $H(t) = H(t + \Omega)$ will be named **linear periodic γ-Hamiltonian system**, where $x \in \mathbb{R}^{2n}$ and $H^T(t) = H(t)$ are a $2n \times 2n$ matrix and $\gamma \ge 0$.

Remark 29 *According to Lemma 18, the state transition matrix* $\Phi(t, t_0)$ *of Eq. (33) is* μ*-symplectic, in particular, the state transition matrix evaluated over one period* Ω*.*

Corollary 30 *The* ***monodromy matrix*** $M = e^{R\Omega}$ *and the matrix R of the periodic system in Eq. (33) are* μ*-symplectic and* γ*-Hamiltonian matrices, respectively, with* $\mu = e^{-2\gamma\Omega}$.

Proof 31 *From the definition of a* μ*-symplectic matrix* $M^TJM = \left(e^{R\Omega}\right)^T J\left(e^{R\Omega}\right) = \mu J$, *we obtain* $e^{R^T\Omega} = \mu J e^{-R\Omega} J^{-1} = \mu J \left\{ I_{2n} - R\Omega + \frac{RR\Omega^2}{2} - \frac{RRR\Omega^3}{3!} + \ldots + \frac{R^k\Omega^4}{k!} + \ldots \right\} J^{-1}$ $= \mu e^{-JRJ^{-1}\Omega} = e^{-2\gamma\Omega} e^{-JRJ^{-1}\Omega}$ *thus* $e^{R^T\Omega} = e^{-2\gamma\Omega - JRJ^{-1}\Omega} \Rightarrow R^TJ + JR = -2\gamma J$.

This corollary states the main relation in our analysis. The symmetry of the μ-symplectic matrix will be utilized for obtaining the stability conditions of the system in Eq. (33). Furthermore, by applying the Lyapunov transformation

$$z(t) = P(t)x(t) \tag{34}$$

we conclude that any linear periodic γ-Hamiltonian system can be reduced to a linear time-invariant γ-Hamiltonian system

$$\dot{z}(t) = Rz(t). \tag{35}$$

The next two subsections are based on [12] and are adapted for characteristic polynomials of μ-symplectic matrices.

4.1 Stability of a system with one degree of freedom

For $n = 1$, the characteristic polynomial of the monodromy matrix M associated with the system in Eq. (33) becomes $P_M(\lambda) = \lambda^2 + a\lambda + \mu$ with $a = -\text{tr}(M)$. According to the Lemma 18, M is μ-symplectic. Then, there are two multipliers symmetric to the circle of radius r and the real axis. Therefore, the multipliers only can leave the unit circle at the coordinates $(1, 0)$ or $(-1, 0)$ (see **Figure 2**). Note that the term $-a$ is equal to the transformation in Eq. (13):

$$\delta = \lambda + \frac{\mu}{\lambda} = \text{tr}(M) = -a$$

Theorem 32 *For $n = 1$, the system in Eq. (33) is asymptotically stable if and only if the inequality*

$$|a| < (1 + \mu)$$

is satisfied.

Proof 33 *Since the multipliers only leave the unit circle on the points $\lambda = 1$ or $\lambda = -1$, the stability boundaries are given by*

$$\begin{aligned} P_M(1) &= (1)^2 + a(1) + \mu = a + (\mu + 1) \\ P_M(-1) &= (-1)^2 + a(-1) + \mu = -a + (\mu + 1) \end{aligned}.$$

This means that $a + (\mu + 1) > 0$ and $-a + (\mu + 1) > 0$ must be fulfilled; thus, $|a| < (1 + \mu)$.

4.2 Stability of a system with two degrees of freedom

For $n = 2$, the characteristic polynomial of the monodromy matrix M associated with the system in Eq. (33) reads

$$P_M(\lambda) = \lambda^4 + a\lambda^3 + b\lambda^2 + a\mu\lambda + \mu^2 \tag{36}$$

where $a = -\text{tr}(M)$ and $2b = (\text{tr}(M))^2 - \text{tr}(M^2)$. There are four multipliers, and due to the symmetry with respect to the μ-circle, they can be categorized in the position configurations depicted in **Figure 2**.

Respecting that the characteristic polynomial is associated with a μ-symplectic matrix, we can use the transformation

$$\delta = \lambda + \frac{\mu}{\lambda} \tag{37}$$

to obtain the auxiliary polynomial

$$Q_M(\delta) = \delta^2 + a\delta + b - 2\mu. \tag{38}$$

The symmetry of the eigenvalues yield

$$a = -\mathrm{tr}(M) = \lambda_1 + \frac{\mu}{\lambda_1} + \lambda_3 + \frac{\mu}{\lambda_3} = \delta_1 + \delta_2.$$

The transition boundaries are characterized by having at least one eigenvalue at $|\lambda| = 1$. The simplest cases are if $\lambda = 1$ ($\delta = 1 + \mu$) or $\lambda = -1$ ($\delta = -1 - \mu$). These points overlay if a real-valued multiplier leaves the unit circle at the point $(1, 0)$ or $(0, -1)$ (see the cases c, d, e, f, or g in **Figure 2**). Substituting these two values into Eq. (36) gives

$$b = -a(1 + \mu) - \left(1 + \mu^2\right) \tag{39}$$

and

$$b = a(1 + \mu) - \left(1 + \mu^2\right). \tag{40}$$

Considering the case $\lambda \in \mathbb{C}$, we search the transition boundary line when two complex multipliers leave the unit circle at points different to $(1, 0)$ and $(0, -1)$ (see cases a or b in **Figure 2**). Then the transition boundary line can be obtained by considering the symmetry of the multipliers with respect to the real axis and the circle of the radius $r = \sqrt{\mu}$. Here, $\lambda_1 = x + iy$, $\lambda_2 = \frac{\mu}{\lambda_1}$, $\lambda_3 = x - iy$, and $\lambda_4 = \frac{\mu}{\lambda_3}$. At $|\lambda| = 1$ $\left(\sqrt{x^2 + y^2} = 1\right)$, it follows that

$$\lambda_1 = x + iy, \quad \lambda_2 = \mu(x - iy), \quad \lambda_3 = y - iy, \quad \lambda_4 = \mu(x + iy).$$

Hence, the transformation in Eq. (13) follows:

$$\delta_1 = \lambda_1 + \frac{\mu}{\lambda_1} = x(1 + \mu) + iy(1 - \mu)$$
$$\delta_2 = \lambda_3 + \frac{\mu}{\lambda_3} = x(1 + \mu) - iy(1 - \mu)$$

Adding δ_1 and δ_2 gives

$$\delta_1 + \delta_2 = 2x(1 + \mu). \tag{41}$$

From Eq. (38) we obtain

$$\delta_{1,2} = \frac{-a}{2} \pm \frac{\sqrt{a^2 + 8\mu - 4b}}{2}. \tag{42}$$

Note that for δ_1 and δ_2 to become complex, the inequality

$$4b > a^2 + 8\mu$$

must be fulfilled. Adding δ_1 and δ_2, one obtains

$$\delta_1 + \delta_2 = -a \tag{43}$$

Equating Eqs. (41) and (43) yields

$$2x(1 + \mu) = -a. \tag{44}$$

The real part x of the eigenvalues results from Eq. (37)

$$\lambda^2 - \lambda\delta + \mu = 0$$

and

$$\lambda = \frac{\delta \pm \sqrt{\delta^2 - 4\mu}}{2}. \tag{45}$$

Substituting Eq. (42) into Eq. (45) and choosing only the positive signs gives

$$\lambda_1 = \frac{1}{4}\left(-a + \sqrt{w - 4\mu - 2b + a^2}\right) + \frac{i}{4}\left(\sqrt{w + 4\mu + 2b - a^2} + \sqrt{-(a^2 + 8\mu) + 4b}\right)$$

with the abbreviation $w = 2\sqrt{-4a^2\mu + (b + 2\mu)^2}$. Consequently, the real part of λ is

$$x = \frac{1}{4}\left(-a + \sqrt{w - 4\mu - 2b + a^2}\right),$$

and substituting into Eq. (44) results in

$$\frac{1}{2}\left(-a + \sqrt{w - 4\mu - 2b + a^2}\right)(1 + \mu) = -a$$

which can be solved for b to obtain the transition boundary curve

$$b = \frac{\mu^4 + 2\mu^3 + 2\mu^2 + 2\mu + a^2\mu + 1}{(1 + \mu)^2}. \tag{46}$$

Two intersection points exist on each line in Eqs. (39) and (40) with the curve defined by Eq. (46). These points are

$$b = \frac{1}{\mu}\left(\mu^4 + \mu^3 + 2\mu^2 + \mu + 1\right) \tag{47}$$

$$b = \mu^2 + 4\mu + 1 \tag{48}$$

and are highlighted in **Figure 3**. The line in Eq. (47), dashed line in the figure, is a necessary condition for stability.

Theorem 34 *The Eq. (33) when $n = 2$ is asymptotically stable if and only if the inequalities are fulfilled:*

$$b \geq -a(1 + \mu) - \left(1 + \mu^2\right), \tag{49}$$

$$b \geq a(1 + \mu) - \left(1 + \mu^2\right), \tag{50}$$

$$b \leq \frac{\mu^4 + 2\mu^3 + 2\mu^2 + a^2\mu + 2\mu + 1}{(1 + \mu)^2}. \tag{51}$$

From this analysis, the multipliers position in relation to the unit circle and μ-circle are defined by inequalities. These split the complex plane into four regions as it is shown in the **Figure 3**.

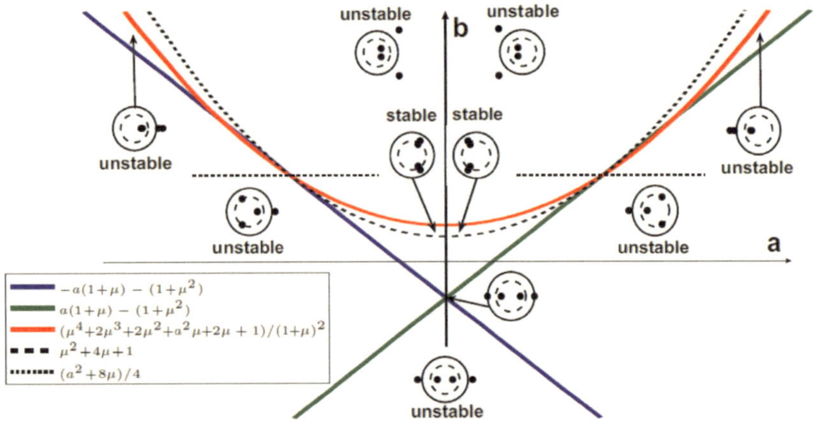

Figure 3.
Multiplier map in the case of $n = 2$: Horizontal and vertical axes are the coefficients a and b of the characteristic polynomial of the monodromy matrix M in Eq. (36). The solid lines represent the borders of the inequalities in Theorem 34, Eqs. (49), (50), and (51). The dots indicate the position of multipliers and the unit circle, in solid line, associated with the system in Eq. (33) in the case of $n = 2$. The dashed circle depicts the μ-circle.

5. Coupled Mathieu equations

Consider two coupled and damped Mathieu equations of the following form:

$$\begin{bmatrix} \ddot{z}_1 \\ \ddot{z}_2 \end{bmatrix} + \begin{bmatrix} \Theta_{11} & \Theta_{12} \\ \Theta_{21} & \Theta_{22} \end{bmatrix} \begin{bmatrix} \dot{z}_1 \\ \dot{z}_2 \end{bmatrix} + \left(\begin{bmatrix} \omega_1^2 & 0 \\ 0 & \omega_2^2 \end{bmatrix} + \beta \begin{bmatrix} Q_{11} & Q_{12} \\ Q_{21} & Q_{22} \end{bmatrix} \begin{bmatrix} z_1 \\ z_2 \end{bmatrix} \cos(\nu t) \right) \begin{bmatrix} z_1 \\ z_2 \end{bmatrix} = 0. \tag{52}$$

Following the procedure presented in Section 3.1, the system in Eq. (52) can be cast into the γ-Hamiltonian form in Eq. (33) if $\Theta_{12} = \Theta_{21}$ and $Q_{12} = Q_{21}$, i.e., the coefficient matrices Θ and Q are symmetric. In this case, the coupled Mathieu equations present all the properties of the periodic γ-Hamiltonian system defined in Eq. (33) for $n = 2$ and $\Omega = 2\pi/\nu$. Hence, all the above analysis on Hamiltonian systems can be applied. The monodromy matrix is computed by numerical methods, and the stability chart is obtained by applying the Theorem 34.

The following numerical values are chosen for the analysis of a specific system $\omega_1^2 = 8$, $\omega_2^2 = 26$, $Q_{11} = Q_{22} = 2$, $Q_{12} = Q_{21} = -2$. **Figure 4a** depicts the multiplier chart similar to **Figure 3**. The unstable regions are colored and the stable regions are kept white. Each color depicts a specific configuration of the multiplier positions within the unit circle and the μ-circle according to the inequalities stated in Theorem 34 and visualized in **Figures 3** and **4a**. The description of each color is relevant because each color describes the parametric resonance phenomenon. Thus, yellow, magenta, and cyan colors refer to the configuration of four real-valued multipliers, two of them inside and two outside of the unit circle. These multipliers are either all negative (magenta region), all positive (yellow region), or two positive and two negative (cyan region). The blue and red regions indicate two complex conjugate multipliers on the μ-circle, while the other two are real with $|\lambda| > 1$. The two real multipliers are either positive (blue) or negative (red). Then, all four multipliers are complex conjugate within the green region. In this case, two multipliers lie inside and two outside of the unit circle.

DOI: http://dx.doi.org/10.5772/intechopen.88635

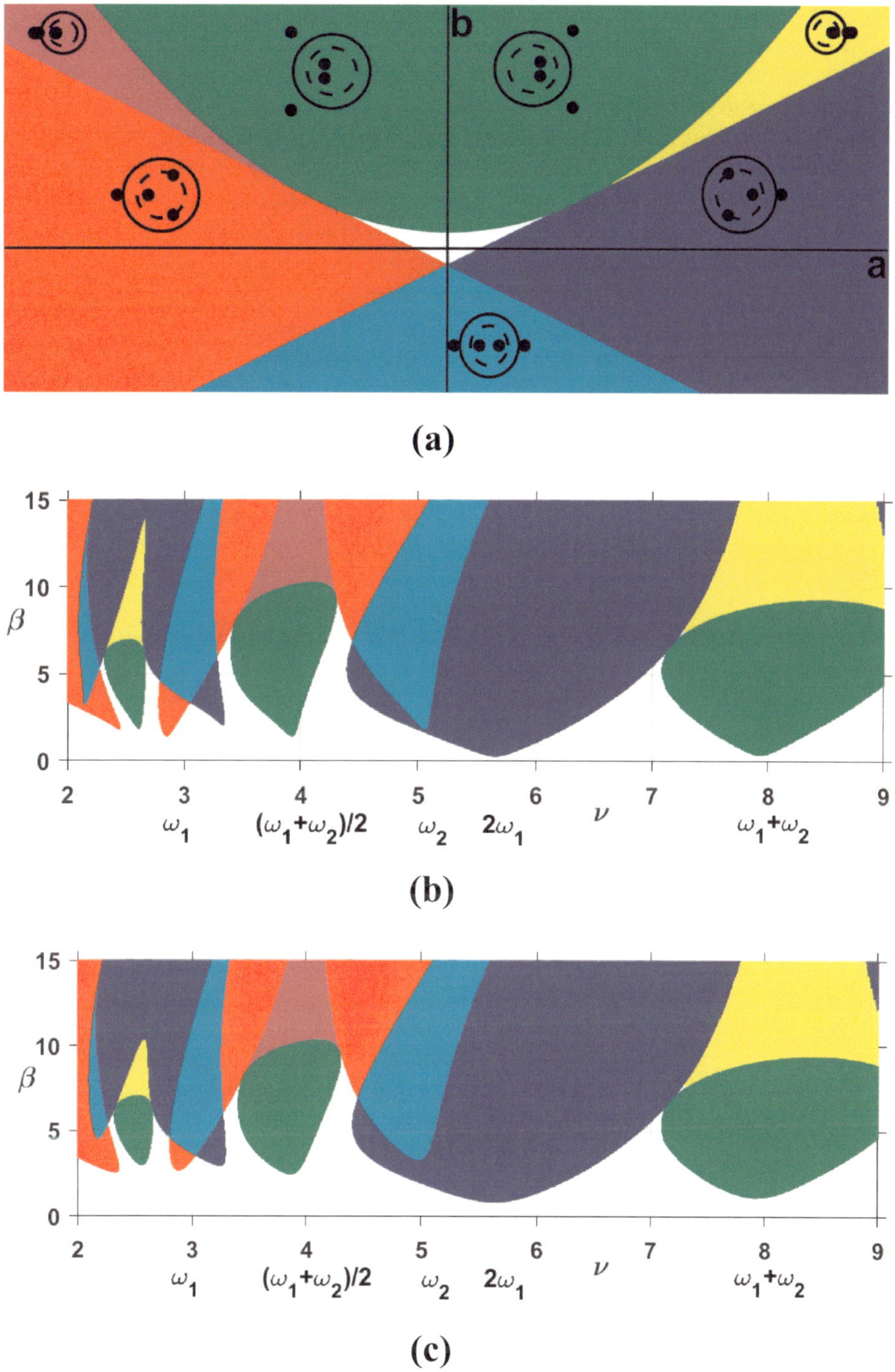

Figure 4.
Multiplier map and stability charts, for example, systems in Eq. (52). Multiplier map corresponds to ***Figure 3*** *but now with colored regions for the different unstable multiplier configurations. Stability charts are given for different values of damping. (a) Multiplier map: a and b are the coefficients of the corresponding characteristic polynomial of the monodromy matrix. All colored zones correspond to unstable positions multipliers configurations. (b) Stability chart of coupled Mathieu equations, Eq. (52), with small damping:* $\Theta_{12} = \Theta_{21} = 0$ *and* $\Theta_{11} = \Theta_{22} = 0.1$*. Each color code is according to the position of the multipliers as in* ***Figure 4a****. (c) Stability chart of coupled Mathieu equations, Eq. (52), with damping:* $\Theta_{12} = \Theta_{21} = 0$ *and* $\Theta_{11} = \Theta_{22} = 0.3$*.*

Additionally, parametric primary resonances occur at parametric excitation frequencies $\nu = 2\omega_i/k$, with $k \in \mathbb{N}^+$, and parametric combination resonances of summation type occur at $\nu = (\omega_1 + \omega_2)/k$ [7, 10]. These frequencies are also observed for the example system in **Figure 4**. The green regions mark parametric combination resonances. The blue and red regions correspond to parametric primary resonances. The presented calculation technique can be categorized as a semi-analytical method. After rewriting the original system into the form in Eq. (33), the monodromy matrix is constructed by integrating the equations of motion using numerical methods.

Subsequently, the coefficients of the characteristic polynomial of the monodromy matrix can be computed as $a = -\mathrm{tr}(M)$ and $2b = (\mathrm{tr}(M))^2 - \mathrm{tr}(M^2)$. This technique **avoids the computation of the eigenvalues itself**. This has the main advantage that numerical problems on the computation of the eigenvalues are avoided, e.g., numerical sensitivity of multipliers [21].

The definitions of μ-symplectic and γ-Hamiltonian matrices allow the analysis of a linear periodic Hamiltonian system with a particular dissipation. The main result of the proposed theory lies in Corollary 30 which states that the state transition matrix of any γ-Hamiltonian system is μ-symplectic. The symmetry properties of the eigenvalues of μ-symplectic matrices lead to an efficient calculation of the stability boundaries of this type of system. The general framework is applied for the example analysis of two damped and coupled Mathieu equations confirming the faster and robust computation of the stability chart. The procedure can be extended to a higher number of coupled Mathieu equations as outlined above.

Acknowledgements

M.Ramírez is thankful to Conacyt for the provided support during his Ph.D. studies in México and to the Austrian EFRE/LEADER project LaZu-CLLD IWB TIROL OSTT 005 for his research stay at Campus Technik Lienz.

Appendix

Proof 35 *In* [22], *Rim proposed an elementary proof that real symplectic matrices have determinant one; following the same procedure, we prove that the symplectic matrices have determinant* μ^n. *From the definition* $\det(M^T J M) = \det(M^T)\det(J)\det(M) = \det(\mu J) = \mu^{2n} \rightarrow \det(M) = \pm\mu^n$, *therefore it is necessary to prove that* $\det[M] = -\mu^{2n}$ *is false. Considering the matrix* $S = M^T M + \mu I_{2n}$ *since* $M^T M \geq 0$ *and* $\mu \in (0, 1]$, *the matrix S has real and greater than* μ *eigenvalues:*

$$\det(S) = \det\left(M^T M + \mu I_{2n}\right) > \mu. \tag{53}$$

Now from the definition $M^{-T} = \mu^{-1} J M J^{-1}$ *and rewriting S,*

$$S = M^T M + \mu I_{2n} = M^T\left(M + \mu M^{-T}\right) = M^T\left(M + JMJ^{-1}\right)$$

denotes the subblocks of M as follows:

$$M = \begin{bmatrix} M_{11} & M_{12} \\ M_{21} & M_{21} \end{bmatrix}$$

with $M_{11}, M_{12}, M_{21}, M_{21} \in \mathbb{R}^{n\times n}$; thus $M + JMJ^{-1} = \begin{bmatrix} M_{11}+M_{22} & M_{12}-M_{21} \\ M_{21}-M_{12} & M_{11}+M_{22} \end{bmatrix}$ if $A = M_{11} + M_{22}$ and $B = M_{12} - M_{21}$; then

$$M + JMJ^{-1} = \begin{bmatrix} A & B \\ -B & A \end{bmatrix}. \tag{54}$$

We rewrite (54) with the unitary transformation T:

$$T = \frac{1}{\sqrt{2}}\begin{bmatrix} I_n & I_n \\ iI_n & -iI_n \end{bmatrix} \Leftrightarrow T^{-1} = \frac{1}{\sqrt{2}}\begin{bmatrix} I_n & -iI_n \\ I_n & iI_n \end{bmatrix}$$

$$M + JMJ^{-1} = \begin{bmatrix} A & B \\ -B & A \end{bmatrix} = \frac{1}{\sqrt{2}}\begin{bmatrix} I_n & I_n \\ iI_n & -iI_n \end{bmatrix}\begin{bmatrix} A+iB & 0 \\ 0 & A-iB \end{bmatrix}\frac{1}{\sqrt{2}}\begin{bmatrix} I_n & -iI_n \\ I_n & iI_n \end{bmatrix}$$

therefore

$$\begin{aligned} \det(S) &= \det\left(M^T\left(M + JMJ^{-1}\right)\right) = \det\left(M^T\right)\det\left(M + JMJ^{-1}\right) \\ &= \det(M)\det(A+iB)\det(A-iB) = \det(M)\det(A+iB)\det\left(\overline{A+iB}\right) \end{aligned}$$

since *A*, *B* are real, the complex conjugation commute with the determinant, and then

$$\det(S) = \det(M)\det(A+iB)\overline{\det(A-iB)} = \det(M)|\det(A+iB)|^2$$

and from Eq. (53)

$$0 < \mu < \det(S) = \det(M)|\det(A+iB)|^2$$

then $|\det(A+iB)|^2 > 0$ *and* $\det(M) > 0$; *therefore* $\det(M) = \mu^n$. ■

Author details

Miguel Ramírez Barrios[1*], Joaquín Collado[2] and Fadi Dohnal[3]

1 Professional Interdisciplinary Unit of Biotechnology, National Polytechnic Institute, Mexico

2 Automatic Control Department, Cinvestav, Mexico City, Mexico

3 Division for Mechatronics Lienz, UMIT, Lienz, Austria

*Address all correspondence to: mramirez@ctrl.cinvestav.mx
and fadi.dohnal@umit.at

References

[1] Sinha SC, Butcher E, Dávid A. Construction of dynamically equivalent time-invariant forms for time-periodic systems. Nonlinear Dynamics. 1998;**16**(3):203-221. DOI: 10.1023/A:1008072713385

[2] Butcher E, Sinha SC. Normal forms and the structure of resonance sets in nonlinear time-periodic systems. Nonlinear Dynamics. 2000;**23**(1):35-55. DOI: 10.1023/A:1008312424551

[3] Ng L, Rand R. Bifurcations in a Mathieu equation with cubic nonlinearities. Chaos, Solitons & Fractals. 2002;**14**(2):173-181. DOI: 10.1016/S0960-0779(01)00226-0

[4] Cartmell M. Introduction to Linear, Parametric and Nonlinear Vibrations. London: Chapman and Hall; 1990

[5] Biswas S, Bhattacharjee J. On the properties of a class of higher-order Mathieu equations originating from a parametric quantum oscillator. Nonlinear Dynamics. 2009;**96**(1):737-750. DOI: 10.1007/s11071-019-04818-9

[6] Hansen L. Stability diagrams for coupled Mathieu-equations. Archive of Applied Mechanics. 1985;**55**(6):463-473. DOI: 10.1007/BF00537654

[7] Dohnal F, Verhulst F. Averaging in vibration suppression by parametric stiffness excitation. Nonlinear Dynamics. 2008;**54**(3):231-248. DOI: 10.1007/s11071-007-9325-z

[8] Collado J. Hill equation from 1 to 2 degrees of freedom. In: New Perspectives and Applications of Modern Control Theory. Switzerland: Springer Nature; 2018. pp. 43-71. DOI: 10.1007/9783319624648_3

[9] Meyer KR, Hall GR, Offin D. Introduction to Hamiltonian Dynamical Systems and the N-Body Problem. New York: Springer; 2009. DOI: 10.1007/978-3-319-53691-0

[10] Yakubovich VA, Starzhinskii VM. Linear Differential Equations With Periodic Coefficients. New York: John Wiley and Sons; 1975

[11] Dragt A. Lie methods for nonlinear dynamics with applications to accelerator physics. Physics Department Report. University of Maryland; 2011

[12] Howard JE, MacKay RS. Linear stability of symplectic maps. Journal of Mathematical Physics. 1987;**28**(5):1036-1051. DOI: 10.1063/1.527544

[13] Ramírez M, Collado J, Dohnal F. Stability of coupled and damped Mathieu equations utilizing symplectic properties. In: Nonlinear Dynamics of Structures, Systems and Devices of the Proceedings of the First International Nonlinear Dynamics Conference (NODYCON2019)

[14] Seyranian AP, Mailybaev AA. Multiparameter Stability, Theory with Applications. USA: World Scientific; 2003. DOI: 10.1142/5305

[15] Horn RA, Johnson CR. Topics in Matrix Analysis. New York: Cambridge University Press; 1991. DOI: 10.1017/CBO9780511840371

[16] Adrianova LY. Introduction to linear systems of differential equations. In: Translations of Mathematical Monographs. Vol. 146. Providence, Rhode Island: American Mathematical Society; 1995

[17] Awrejcewicz J. Ordinary Differential Equations and Mechanical Systems. Switzerland: Springer International Publishing; 2014. DOI: 10.1007/978-3-319-07659-1

[18] Sinha SC, Pandiyan RR, Bibb JS. Liapunov-Floquet transformation:

Computation and applications to periodic systems. ASME. Journal of Vibration and Acoustics. 1996;**118**(2): 209-219. DOI: 10.1115/1.2889651

[19] Richards JA. Analysis of Periodically Time-Varying Systems. Berlin: Springer-Verlag; 1983. DOI: 10.1007/978-3-642-81873-8

[20] Yakubovich VA. A remark on the Floquet-Lyapunov theorem. Vestnik Leningrad University. 1970;**25**(1):88-92

[21] Ramírez M, Collado J. Calculation of the stability zones of Hill's equation with a GPU on Matlab. In: International Conference on Supercomputing; Springer; 2015. pp. 225-239. DOI: 10.1007/978-3-319-32243-8_16

[22] Rim D. An elementary proof that symplectic matrices have determinant one. Advances in Dynamical Systems and Applications (ADSA). 2017;**12**(1): 15-20. arXiv:1505.04240

Chapter 2

Partial Entropy and Bundle-Like Entropy for Topological Dynamical Systems

Kesong Yan and Fanping Zeng

Abstract

Entropy is an important notion for understanding the complexity of dynamical systems. Several important entropy-like invariants based on the preimage structure for noninvertible maps have been defined and studied by some authors. In this chapter, following the idea of Hurley, we first further study the relationship among the topological entropy, pseudo-orbit, and preimage entropies for topological dynamical systems from the view of localization. Secondly, two entropy-like invariants, which are called the partial entropy and bundle-like entropy, for nonautonomous discrete dynamical systems are introduced. A relationship between the topological entropy and such two entropies is established.

Keywords: topological entropy, point entropy, pseudo-orbit, partial entropy, bundle-like entropy
2000 Mathematics Subject Classification: Primary: 37B40, 37A35, 37B10, 37A05

1. Introduction

By a *topological dynamical system*, we mean a pair (X, T), where X is a compact metric space with a metric d and T is a continuous surjective map from X to itself [1]. An important notion for understanding the complexity of dynamical systems is topological entropy, which was first introduced by Adler et al. [2] in 1965, and later Dinaburg [3] and Bowen [4] gave two equivalent definitions on a metric space by using separated sets and spanning sets. Roughly speaking, topological entropy measures the maximal exponential growth rate of orbits for an arbitrary topological dynamical system.

When a considered mapping T is invertible, it is well-known that T and the inverse mapping T^{-1} have the same topological entropy. However, when the map T is not invertible, the "inverse" is set-valued, yielding the iterated preimage set $T^{-n}(x) = \{z \in X : T^n z = x\}$ of a point $x \in X$ which is in general a set rather than a point, so different ways of "extending the procedure into the past" lead to several new entropy-like invariants for non-invertible maps.

In 1991, Langevin and Walczak [5] regard the "inverse" as a relation and formulate a notion of entropy for this relation (analogous to the entropy of a foliation [6]), based on distinguishing points by means of the structure of their "preimage trees," which is called preimage relation entropy. The interested reader can see [7] or [8] for more details on this invariant. Later, several important entropy-like invariants based on the preimage structure for non-invertible maps, such as pointwise preimage entropies, preimage branch entropy [1, 8–10], partial preimage entropy, conditional preimage entropy [11], etc., have been introduced, and their relationships with topological entropy have been established. To learn more about the results related to the preimage entropy for noninvertible maps, one can see [12–23].

The local entropy theory for topological dynamical systems started in the early 1990s with the work of Blanchard (see [24, 25]). Nowadays this theory has become a very interesting topic in the field of dynamical systems and has also proven to be fundamental to many other related fields. For example, Blanchard defined the notion of entropy pairs and used it to obtain a disjointness theorem [26]. The notion of entropy pairs can also be used to show the existence of the maximal zero-entropy factor, called the topological Pinsker factor, for any topological dynamical system [25]. In order to gain a better understanding of the topological version of a K-system, the theory of entropy tuples [27–29] was developed. To learn more about the theory related to the local entropy, we refer the interested reader to see the survey paper [30] and references therein.

We remark that in reality, it is difficult to find a real orbit in the system, but a pseudo-orbit can be used to approximate the real orbit, and so there have been a lot of applications in many fields. Since the works of Bowen [31] and Conley [32], pseudo-orbits have proved to be a powerful tool in dynamical systems. For instance, Hammel et al. [33, 34] have investigated the role of pseudo-orbits in computer simulations of certain dynamical systems; Barge and Swanson [35] made use of pseudo-orbits to study rotation sets of circle and annulus maps. Also, a remarkable result by Misiurewicz [36] showed that the topological entropy can be computed by measuring the exponential growth rate of the numbers of pseudo-orbits (related results can see [37]). In [1], Hurley showed that the point entropy with pseudo-orbits that is defined by replacing inverse orbit segments by inverse pseudo-orbit segments in the definition of pointwise preimage entropy is in fact equal to the topological entropy.

In this chapter, following Hurley [1] we further study the preimage entropy for topological dynamical system from the view of localization. In Section 2, we consider the calculation of topological entropy for open covers from pseudo-orbits (Theorem 2.3). In Section 3, we investigate the relationship among the topological entropy for open covers and several preimage entropy invariants, which is viewed as the local version of the Hurley inequality (Theorem 3.1). In Section 4, we show that the topological entropy for open covers can be computed by measuring the exponential growth rate of the number of pseudo-orbits that end at a particular point (Theorems 4.2 and 4.3).

A nonautonomous discrete dynamical system is a natural generalization of a classical dynamical system; its dynamics is determined by a sequence of continuous self-maps $f_n : X_n \to X_{n+1}$, which defined on a sequence on compact metric spaces (X_n, d_n). The topological entropy of nonautonomous discrete dynamical systems was introduced by Kolyada and Snoha [38]. In Section 5, following the idea of [1, 39], we introduce two entropy-like invariants, which are called the partial entropy and bundle-like entropy, for nonautonomous discrete dynamical systems, and study the relationship among them and the topological entropy (Theorems 5.2, 5.3, and 5.5).

2. Topological entropy and pseudo-orbits

2.1 Topological entropy via open covers

Topological entropy was defined originally by Adler et al. [2] for continuous maps on compact topological spaces. Let (X, T) be a topological dynamical system. A finite *open cover* of X is a finite family of open sets whose union is X. Denoted by

$\mathcal{C}_X^o$ is the set of finite open covers of X. Given two open covers $\mathcal{U}, \mathcal{V} \in \mathcal{C}_X^o$, $\mathcal{U}$ is said to be *finer than* $\mathcal{V}$ ($\mathcal{U} \succcurlyeq \mathcal{V}$) if each element of $\mathcal{U}$ is contained in some element of $\mathcal{V}$. Let $\mathcal{U} \vee \mathcal{V} = \{U \cap V : U \in \mathcal{U}, V \in \mathcal{V}\}$. It is clear that $\mathcal{U} \vee \mathcal{V} \succcurlyeq \mathcal{U}$ and $\mathcal{U} \vee \mathcal{V} \succcurlyeq \mathcal{V}$.

Let $\mathcal{U} \in \mathcal{C}_X^o$. For two nonnegative integers $M \le N$, denoted by $\mathcal{U}_M^N = \vee_{n=M}^N T^{-n}\mathcal{U}$, where $T^{-n}\mathcal{U} = \{T^{-n}(U) : U \in \mathcal{U}\}$ for all positive integers n. For any $K \subset X$, define $N(\mathcal{U}|K)$ as the minimal cardinality of any subcovers of $\mathcal{U}$ that covers K. We write $N(\mathcal{U}|X)$ simply by $N(\mathcal{U})$. The *topological entropy of* $\mathcal{U}$ *with respect to* T is defined by

$$h_{\text{top}}(T, \mathcal{U}) = \lim_{n\to\infty} \frac{1}{n} \log N\left(\mathcal{U}_0^{n-1}\right) = \inf_{n\ge 1} \frac{1}{n} \log N\left(\mathcal{U}_0^{n-1}\right). \tag{1}$$

The *topological entropy of* T is

$$h_{\text{top}}(T) = \sup_{\mathcal{U} \in \mathcal{C}_X^o} h_{\text{top}}(T, \mathcal{U}). \tag{2}$$

2.2 Separated sets, spanning sets, and topological entropy

In this subsection, we recall two equivalent definitions, which are given by Dinaburg [3] and Bowen [4]. Let (X, T) be a topological dynamical system. Given a nonempty subset K of X, for any $\epsilon > 0$ and $n \in \mathbb{N}$, a subset E of K is called an (n, ϵ)-*separated set* of K if any $x \ne y \in E$ implies $d_n(x, y) \ge \epsilon$, where

$$d_n(x, y) := \max_{0 \le i \le n-1} d\left(T^i x, T^i y\right).$$

Denote the maximal cardinality of any (n, ϵ)-separated subset of K by $s(n, \epsilon, K)$. A subset F of K is called an (n, ϵ)-*spanning set* of K, if for any $x \in K$, there exists $y \in F$ with $d_n(x, y) < \epsilon$. Denote the minimal cardinality of any (n, ϵ)-spanning set for K by $r(n, \epsilon, K)$.

The following lemma is well-known, and its proof is not difficult, so we omit its detail proof.

Lemma 2.1. *Let* (X, T) *be a topological dynamical system. For any subset K of X and any integer* $n \ge 1$, *we have the following properties*:

1. $r(n, \epsilon, K) \le s(n, \epsilon, K) \le r(n, \epsilon/2, K)$ *for all* $\epsilon > 0$.

2. $N\left(\mathcal{U}_0^{n-1}|K\right) \le r(n, \delta, K)$ *for any* $n \in \mathbb{N}$ *and any* $\mathcal{U} \in \mathcal{C}_X^o$ *with the Lebesgue number* 2δ.

3. $s(n, \epsilon, K) \le N\left(\mathcal{U}_0^{n-1}|K\right)$ *for any* $\mathcal{U} \in \mathcal{C}_X^o$ *with* $\text{diam}(\mathcal{U}) < \epsilon$.

By Lemma 2.1, we obtain directly the following result.

Theorem 2.2. (see [3, 4, 40]). *Let* (X, T) *be a topological dynamical system. Then*

$$h_{\text{top}}(T) = \lim_{\epsilon\to 0} \limsup_{n\to\infty} \frac{1}{n} \log s(n, \epsilon, X) = \lim_{\epsilon\to 0} \limsup_{n\to\infty} \frac{1}{n} \log r(n, \epsilon, X).$$

2.3 Topological entropy via pseudo-orbits

Let (X, d) be a compact metric space. Denote X^n as the n-fold Cartesian product of X $(n \ge 1)$. Fixing a positive number ϵ, a subset $E \subset X^n$ is said to be (n, ϵ)-*separated* if for any two distinct points $\tilde{x} = (x_0, x_1, \cdots, x_{n-1}), \tilde{y} = (y_0, y_1, \cdots, y_{n-1}) \in E$, there is a $0 \le i \le n-1$ such that $d(x_i, y_i) > \epsilon$. By the compactness of X, any (n, ϵ)-separated set is finite. If $Z \subset X^n$ is a nonempty subset, then we denote the maximal cardinality of any (n, ϵ)-separated subset of Z by $s(n, \epsilon, Z)$.

Let $Z \subset X^n$ be a nonempty subset. A subset $F \subset Z$ is called (n, ϵ)-*panning* for Z if for each $\tilde{z} = (z_0, z_1, \cdots, z_{n-1}) \in Z$, there is a $\tilde{y} = (y_0, y_1, \cdots, y_{n-1}) \in F$ with $d(z_i, y_i) < \epsilon$ for every $0 \le i \le n-1$. We denote the minimal cardinality of any (n, ϵ)-spanning subset of Z by $r(n, \epsilon, Z)$.

For each positive integer $n \ge 1$, we let O_n denote the set of all orbit segments of length n, that is,

$$O_n = \left\{ \left(x, Tx, \cdots, T^{n-1}x\right) \in X^n : x \in X \right\}.$$

Note that a point $\tilde{w} = \left(x, Tx, \cdots, T^{n-1}x\right) \in O_n$ is uniquely determined by its initial point $x \in X$. Thus, we have

$$\begin{aligned} h_{\text{top}}(T) &= \lim_{\epsilon \to 0} \lim_{\alpha \to 0} \limsup_{n \to \infty} \frac{1}{n} \log s(n, \epsilon, O_n) \\ &= \lim_{\epsilon \to 0} \lim_{\alpha \to 0} \limsup_{n \to \infty} \frac{1}{n} \log r(n, \epsilon, O_n). \end{aligned}$$

Topological entropy has been characterized by Misiurewicz [36] and Barge and Swanson [37] in terms of growth rates of pseudo-orbits. Let (X, T) be a topological dynamical system. For $\alpha > 0$, an *α-pseudo-orbit* for T of length n is a point $\tilde{x} = (x_0, x_1, \cdots, x_{n-1}) \in X^n$ with the property that $d\left(T\left(x_{j-1}\right), x_j\right) < \alpha$ for all $1 \le j \le n-1$. Let $\Psi_n(\alpha) \subset X^n$ denote all α-pseudo-orbits of length n. It was shown in [36, 37] that

$$\begin{aligned} h_{\text{top}}(T) &= \lim_{\epsilon \to 0} \lim_{\alpha \to 0} \limsup_{n \to \infty} \frac{1}{n} \log s(n, \epsilon, \Psi_n(\alpha)) \\ &= \lim_{\epsilon \to 0} \lim_{\alpha \to 0} \limsup_{n \to \infty} \frac{1}{n} \log r(n, \epsilon, \Psi_n(\alpha)). \end{aligned}$$

In the following, we will show that the topological entropy for an open cover can be characterized by pseudo-orbits. Before proceeding, let us first introduce a definition of pseudo-orbit entropy via open covers. Let (X, T) be a topological dynamical system. For each integer $n \ge 1$ and $\mathcal{U} \in \mathcal{C}_X^o$, we define an open cover $\mathcal{U}^n$ of the product space X^n by

$$\mathcal{U}^n = \left\{ U_0 \times U_1 \times \cdots \times U_{n-1} : U_j \in \mathcal{U} \ \text{ for each } \ j = 0, 1, \ldots, n-1 \right\},$$

where

$$U_0 \times U_1 \times \cdots \times U_{n-1} = \left\{ (u_0, u_1, \ldots, u_{n-1}) : u_j \in U_j \ \text{ for each } \ j = 0, 1, \ldots, n-1 \right\}.$$

Given $\alpha > 0$, it is not hard to see that $a_n = N(\mathcal{U}^n | \Psi_n(\alpha))$ is a nonnegative subadditive sequence, that is, $a_{n+m} \le a_n + a_m$ for all positive integers n and m. The *α-pseudo-orbit entropy* of $\mathcal{U}$ is then defined by

$$h_\Psi(T, \mathcal{U}, \alpha) = \lim_{n \to \infty} \frac{1}{n} \log N(\mathcal{U}^n | \Psi_n(\alpha)) = \inf_{n \ge 1} \frac{1}{n} \log N(\mathcal{U}^n | \Psi_n(\alpha)), \tag{3}$$

and the *pseudo-orbit entropy* of $\mathcal{U}$ is defined by

$$h_\Psi(T, \mathcal{U}) = \lim_{\alpha \to 0} h_\Psi(T, \mathcal{U}, \alpha). \tag{4}$$

Theorem 2.3. *Let* (X, T) *be a topological dynamical system. If* $\mathcal{U} \in \mathcal{C}_X^o$*, then we have*

$$h_{\text{top}}(T, \mathcal{U}) = h_\Psi(T, \mathcal{U}). \tag{5}$$

Proof. To prove (5), it suffices to note that $h_\Psi(T, \mathcal{U}, \alpha_1) \le h_\Psi(T, \mathcal{U}, \alpha_2)$ whenever $\alpha_1 < \alpha_2$ and $\inf_{0<\alpha\le 1} N(\mathcal{U}^n | \Psi_n(\alpha)) = N(\mathcal{U}^n | O_n) = N(\mathcal{U}_0^{n-1})$. Thus, we have

$$\begin{aligned} h_\Psi(T, \mathcal{U}) &= \lim_{\alpha \to 0} h_\Psi(T, \mathcal{U}, \alpha) \\ &= \inf_{0<\alpha\le 0} \inf_{n \ge 1} \frac{1}{n} \log N(\mathcal{U}^n | \Psi_n(\alpha)) \\ &= \inf_{n \ge 1} \inf_{0<\alpha\le 1} \frac{1}{n} \log N(\mathcal{U}^n | \Psi_n(\alpha)) \\ &= \inf_{n \ge 1} \frac{1}{n} \log N(\mathcal{U}_0^{n-1}) = h_{\text{top}}(T, \mathcal{U}). \end{aligned}$$

This completes the proof of the theorem. □

Remark 2.4. Combining (2) and (5), we have

$$h_{\text{top}}(T) = \sup_{\mathcal{U} \in \mathcal{C}_X^o} h_\Psi(T, \mathcal{U}).$$

On the other hand, let us define $h_\Psi(T) = \sup_{\mathcal{U} \in \mathcal{C}_X^o} h_\Psi(T, \mathcal{U})$, which is called the *pseudo-orbit entropy* of T. Using the same techniques of topological entropy (see Lemma 2.1), we can easily show that

$$\begin{aligned} h_\Psi(T) &= \lim_{\epsilon \to 0} \lim_{\alpha \to 0} \limsup_{n \to \infty} \frac{1}{n} \log s(n, \epsilon, \Psi_n(\alpha)) \\ &= \lim_{\epsilon \to 0} \lim_{\alpha \to 0} \limsup_{n \to \infty} \frac{1}{n} \log r(n, \epsilon, \Psi_n(\alpha)). \end{aligned}$$

So, it is in fact to give a simpler proof of Theorem 1 of [37] by Theorem 2.3.

3. Pointwise preimage entropies for open covers and local Hurley's inequality

When T is not invertible, one can ask about growth rates of inverse images $f^{-n}(x)$. In this section we describe two ways of doing this, which were introduced by Hurley in [1].

3.1 Preimage branch entropy

Let (X, T) be a topological dynamical system. Given $x \in X$ let $T_n(x)$ denote the *tree of inverse images of x up to order n*, which is defined by

$$T_n(x) = \left\{ (z_0, z_1, \cdots, z_n) : z_n = x \text{ and } z_j = T(z_{j-1}) \text{ for all } 1 \le j \le n \right\}.$$

Each $(z_0, z_1, \cdots, z_n) \in T_n(x)$ is called a *branch* of $T_n(x)$, and its *length* is n. Note that every branch of $T_n(x)$ ends with x. Let $\mathcal{T}_n = \cup_{x \in X} T_n(x)$; we define a metric on $\mathcal{T}_n$ as follows: suppose that $\tilde{z} = (z_0, z_1, \cdots, z_n)$ and $\tilde{w} = (w_0, w_1, \cdots, w_n)$ are two branches of the length n, the *branch distance* between them is defined as

$$d_{B,n}(\tilde{z}, \tilde{w}) = \max_{0 \leq j \leq n} d(z_j, w_j).$$

Let $O_n = \{T_n(x) : x \in X\}$. Given two trees $T_n(x)$ and $T_n(y)$ in O_n, the *branch Hausdorff* distance between them, $d_{bH}(T_n(x), T_n(y))$ is the usual Hausdorff metric based upon $d_{B,n}$; that is,

$$d_{bH}(T_n(x), T_n(y)) = \max\left\{ \max_{\tilde{z} \in T_n(x)} \min_{\tilde{w} \in T_n(y)} d_{B,n}(\tilde{z}, \tilde{w}), \max_{\tilde{w} \in T_n(y)} \min_{\tilde{z} \in T_n(x)} d_{B,n}(\tilde{z}, \tilde{w}) \right\}.$$

Note that $d_{bH}(T_n(x), T_n(y)) < \epsilon$ if and only if each branch of either tree is $d_{B,n}$ within ϵ of at least one branch of the other tree. Two trees $T_n(x)$ and $T_n(y)$ in O_n are said to be *d_{bH}-(n, ϵ)-separated* if $d_{bH}(T_n(x), T_n(y)) < \epsilon$, that is, there is a branch $\tilde{z}$ of one of the trees with the property that $d_{B,n}(\tilde{z}, \tilde{w}) > \epsilon$ for all branches $\tilde{w}$ of the other tree. Let $t(n, \epsilon)$ denote the maximum cardinality of any d_{bH}-(n, ϵ)-separated sets of O_n. Define the entropy by

$$h_b(T) = \lim_{\epsilon \to 0} \limsup_{n \to \infty} \frac{1}{n} \log t(n, \epsilon),$$

which is called the *preimage branch entropy* of T.

3.2 Pointwise preimage entropies

Let us recall two non-invertible invariants defined by Hurley [1] in 1995. Hurley's invariants are about the maximum rate of dispersal of the preimage sets of individual points, which are called *pointwise preimage entropies* in [8]. The difference between these two invariants is when the maximization takes place:

$$\begin{aligned} h_p(T) &= \sup_{x \in X} \lim_{\epsilon \to 0} \limsup_{n \to \infty} \frac{1}{n} \log s(n, \epsilon, T^{-n}(x)) \\ &= \sup_{x \in X} \lim_{\epsilon \to 0} \limsup_{n \to \infty} \frac{1}{n} \log r(n, \epsilon, T^{-n}(x)), \\ h_m(T) &= \lim_{\epsilon \to 0} \limsup_{n \to \infty} \frac{1}{n} \log \sup_{x \in X} s(n, \epsilon, T^{-n}(x)) \\ &= \lim_{\epsilon \to 0} \limsup_{n \to \infty} \frac{1}{n} \log \sup_{x \in X} r(n, \epsilon, T^{-n}(x)). \end{aligned}$$

It is clear that $h_p(T) \leq h_m(T)$, and in [18] the authors constructed an example for which $h_p(T) < h_m(T)$. In addition, Hurley established the following relationships among preimage branch entropy, pointwise preimage entropy, and topological entropy (see [1], Theorem 3.1):

$$h_m(T) \leq h_{\text{top}}(T) \leq h_m(T) + h_b(T).$$

We call it the Hurley inequality.

3.3 Local Hurley's inequality

In this subsection, we mainly study the relationship among the topological entropy for open covers and several preimage entropy invariants, which is viewed as the local version of the Hurley inequality. To do it, we first introduced a definition of preimage entropy via open covers.

Let (X, T) be a topological dynamical system. Given $\mathcal{U} \in \mathcal{C}_X^o$, define two *pointwise preimage entropies of* $\mathcal{U}$ with respect to T by

$$h_p(T, \mathcal{U}) = \sup_{x \in X} \limsup_{n \to \infty} \frac{1}{n} \log N\left(\mathcal{U}_0^{n-1} | T^{-n}(x)\right)$$

and

$$h_m(T, \mathcal{U}) = \limsup_{n \to \infty} \frac{1}{n} \log \left(\sup_{x \in X} N(\mathcal{U}_0^{n-1} | T^{-n}(x)) \right).$$

Theorem 3.1. (Local Hurley's inequality). *Let* (X, T) *be a topological dynamical system. If* $\mathcal{U} \in \mathcal{C}_X^o$, *then we have*

$$h_p(T, \mathcal{U}) \leq h_m(T, \mathcal{U}) \leq h_{\text{top}}(T, \mathcal{U}) \leq h_m(T, \mathcal{U}) + h_b(T).$$

Proof. It is obvious that $N\left(\mathcal{U}_0^{n-1} | T^{-n}(x)\right) \leq N\left(\mathcal{U}_0^{n-1}\right)$ for every $x \in X$ and every integer $n \geq 1$. So that $h_p(T, \mathcal{U}) \leq h_m(T, \mathcal{U}) \leq h_{\text{top}}(T, \mathcal{U})$. Now we show the last inequality $h_{\text{top}}(T, \mathcal{U}) \leq h_m(T, \mathcal{U}) + h_b(T)$.

Let $\epsilon > 0$ be a Lebesgue number of $\mathcal{U}$. Fix $n \geq 1$, and let Y denote a d_{bH}-$(n, \epsilon/3)$-separated set of O_n with cardinality $t(n, \epsilon/3)$. Let Z denote the set of all root points of trees in Y, where the root point of the tree $T_n(x)$ is x. For each $z \in Z$, let $\mathcal{V}(z, \mathcal{U})$ be a subcover of $\mathcal{U}_0^{n-1}$ with cardinality $N\left(\mathcal{U}_0^{n-1} | T^{-n}(z)\right)$ that covers $T^{-n}(z)$, and let

$$\mathcal{V} = \bigcup_{z \in Z} \mathcal{V}(z, \mathcal{U}).$$

We claim that $\mathcal{V}$ is an open cover of X.

In fact, let $x \in X$ be given and let $w = f^n(x)$. Since Y is a d_{bH}-$(n, \epsilon/3)$-separated set of O_n with maximal cardinality, there is a tree $T_n(y) \in Y$ such that $d_{bH}(T_n(w), T_n(y)) < \epsilon/3$. Now we consider the branch $\tilde{w}$ of $T_n(w)$ begins with x, i.e., $\tilde{w} = \left(x, f(x), \cdots, f^{n-1}(x), f^n(x) = w\right) \in T_n(w)$. Then there exists a branch $\tilde{y} = \left(y_0, y_1, \cdots, y_n = y\right) \in T_n(y)$ such that $d_{B,n}(\tilde{w}, \tilde{y}) < \epsilon/3$. This means that $d\left(T^j(y_0), T^j(x)\right) < \epsilon/3$ for each $0 \leq j \leq n$. Thus, there exists $V \in \mathcal{V}(y, \mathcal{U})$ such that $x \in V$. This yields the claim that $\mathcal{V}$ is an open cover of X. So that $N\left(\mathcal{U}_0^{n-1}\right) \leq |\mathcal{V}|$, where $|\mathcal{V}|$ denotes the cardinality of $\mathcal{V}$. Using the claim, we have

$$\begin{aligned} N\left(\mathcal{U}_0^{n-1}\right) &\leq |\mathcal{V}| \leq \sum_{z \in Z} |\mathcal{V}(z, \mathcal{U})| = \sum_{z \in Z} N\left(\mathcal{U}_0^{n-1} | T^{-n}(z)\right) \\ &\leq |Z| \cdot \left(\sup_{x \in X} N(\mathcal{U}_0^{n-1} | T^{-n}(x)) \right) \\ &= |Y| \cdot \left(\sup_{x \in X} N(\mathcal{U}_0^{n-1} | T^{-n}(x)) \right) = t(n, \epsilon/3) \cdot \left(\sup_{x \in X} N(\mathcal{U}_0^{n-1} | T^{-n}(x)) \right). \end{aligned}$$

So that,

$$\begin{aligned}
h_{\text{top}}(T,\mathcal{U}) &= \lim_{n\to\infty}\frac{1}{n}\log N(\mathcal{U}_0^{n-1}) \\
&\le \limsup_{n\to\infty}\frac{1}{n}\left[\log t(n,\epsilon/3)+\log\left(\sup_{x\in X}N(\mathcal{U}_0^{n-1}|T^{-n}(x))\right)\right] \\
&\le \limsup_{n\to\infty}\frac{1}{n}\log t(n,\epsilon/3)+\limsup_{n\to\infty}\frac{1}{n}\log\left(\sup_{x\in X}N(\mathcal{U}_0^{n-1}|T^{-n}(x))\right) \\
&= \limsup_{n\to\infty}\frac{1}{n}\log t(n,\epsilon/3)+h_m(T,\mathcal{U})\le h_b(T)+h_m(T,\mathcal{U}).
\end{aligned}$$

This completes the proof of the theorem. □

We remark that Theorem 3.1 generalizes the classical Hurley's inequality given in [26, Theorem 3.1]. A direct consequence of Theorem 3.1 is.

Corollary 3.2. (Hurley's inequality). *Let (X, T) be a topological dynamical system. Then we have*

$$h_p(T)\le h_m(T)\le h_{\text{top}}(T)\le h_m(T)+h_b(T). \tag{6}$$

Proof. It follows directly from Lemma 2.1 that

$$h_p(T)=\sup_{\mathcal{U}\in\mathcal{C}_X^o}h_p(T,\mathcal{U})\ \text{ and }\ h_m(T)=\sup_{\mathcal{U}\in\mathcal{C}_X^o}h_p(T,\mathcal{U}). \tag{7}$$

Thus, combining (2), (7), and Theorem 3.1 gives (6). □

4. Point entropy for open covers with pseudo-orbits

In [1], Hurley considered pseudo-orbits for inverse images and showed that the topological entropy can be characterized in terms of growth rates of pseudo-orbits that end at a particular point. Let (X, T) be a topological dynamical system. Recall that if $\alpha>0$, then an α-pseudo-orbit $(x_0, x_1, ..., x_{n-1})\in X^n$ is an approximate orbits segment for T in the sense that $d(T(x_j), x_{j+1})<\alpha$ for all $0\le j\le n-1$.

For each $x\in X$, let $\Psi_n(\alpha,x)\subset X^n$ denote the set of all α-pseudo-orbits of length n that end at x, i.e., an element of $\Psi_n(\alpha,x)$ is an α-pseudo-orbit $(y_0, y_1, \cdots, y_{n-1})$ with $y_{n-1}=x$. It was shown in [1], (Propositions 4.2 and 4.3) that

$$\begin{aligned}
h_{\text{top}} &= \lim_{\epsilon\to0}\lim_{\alpha\to0}\limsup_{n\to\infty}\frac{1}{n}\log\left(\max_{x\in X}s(n,\epsilon,\Psi_n(\alpha,x))\right) \\
&= \sup_{x\in X}\lim_{\epsilon\to0}\lim_{\alpha\to0}\limsup_{n\to\infty}\frac{1}{n}\log s(n,\epsilon,\Psi_n(\alpha,x)).
\end{aligned} \tag{8}$$

In either formula $s(n,\epsilon,\Psi_n(\alpha,x))$ can be replaced by $r(n,\epsilon,\Psi_n(\alpha,x))$.

In the following, we will show that the topological entropy for an open cover can be characterized by pseudo-orbits for inverse images. Before proceeding, let us consider the following definitions, which use the notation introduced in Section 2.3.

Let (X, T) be a topological dynamical system. For each integer $n\ge1$, $\mathcal{U}\in\mathcal{C}_X^o$, and $\alpha>0$, we define

$$N_{\max}(n,\mathcal{U},\alpha)=\max_{x\in X}N(\mathcal{U}^n|\Psi_n(\alpha,x)). \tag{9}$$

Clearly,

$$N(\mathcal{U}^n|\Psi_n(\alpha, x)) \le N_{\max}(n, \mathcal{U}, \alpha) \le N(\mathcal{U}^n|\Psi_n(\alpha)) \tag{10}$$

for every $x \in X$. In addition, by the compactness of X, there is some point $y \in X$ such that

$$N(\mathcal{U}^n|\Psi_n(\alpha, y)) = N_{\max}(n, \mathcal{U}, \alpha).$$

Lemma 4.1. *Let* (X, T) *be a topological dynamical system and* $\mathcal{U} \in \mathcal{C}_X^o$. *Suppose that* $\varepsilon > 0$ *is a Lebesgue number of* $\mathcal{U}$ *and* $0 < \alpha < \varepsilon/4$. *Then there is a constant* $K = K(\alpha)$ *such that for every* $n \ge 1$,

$$N(\mathcal{U}^n|\Psi_n(\alpha)) \le K \cdot N_{\max}(n, \mathcal{U}, \alpha). \tag{11}$$

Proof. Let $\{x_1, x_2, \cdots, x_K\}$ be a finite α-dense subset of X, i.e., $\bigcup_{i=1}^n B(x_i, \alpha) = X$, where $B(x_i, \alpha) = \{z \in X : d(x_i, z) < \alpha\}$. For each $1 \le i \le K$, let $\mathcal{V}_i$ be a subcover of $\mathcal{U}^n$ that covers $\Psi_n(\alpha, x_i)$ with cardinality $N(\mathcal{U}^n|\Psi_n(\alpha, x_i))$. Define $\mathcal{V} = \bigcup_{i=1}^K \mathcal{V}_i$. Clearly, $|\mathcal{V}| \le \sum_{i=1}^K |\mathcal{V}_i| \le K \cdot N_{\max}(n, \mathcal{U}, \alpha)$. So, to complete the proof of the lemma, it suffices to show $\mathcal{V}$ is a subcover of $\mathcal{U}^n$ that covers $\Psi_n(\alpha)$.

In fact, let $\tilde{y} = (y_0, y_1, \cdots, y_{n-1})$ be an α-pseudo-orbit. Since $\{x_1, x_2, \cdots, x_K\}$ is an α-dense subset of X, there is some x_i satisfying $d(T(y_{n-2}), x_i) < \alpha$. This implies $\tilde{z} = (z_0, z_1, \cdots, z_{n-2}, z_{n-1}) = (y_0, y_1, \cdots, y_{n-2}, x_i)$ is an α-pseudo-orbit ending at x_i. Since $\mathcal{V}_i$ is a subcover of $\mathcal{U}^n$ that covers $\Psi_n(\alpha, x_i)$, there is some $V \in \mathcal{V}_i$ such that $\tilde{z} \in V$. Since $z_j = y_j$ for all $0 \le j \le n-2$ and ϵ is the Lebesgue number of $\mathcal{U}$, in order to show that $\tilde{y} \in V$, we need only to show that $d(y_{n-1}, x_i) < \epsilon/2$; this is obviously, as $d(y_{n-1}, x_i) \le d(y_{n-1}, T(y_{n-2})) + d(T(y_{n-2}), x_i) < 2\alpha < \epsilon/2$. □

Theorem 4.2. *Let* (X, T) *be a topological dynamical system. If* $\mathcal{U} \in \mathcal{C}_X^o$, *then we have*

$$h_{\text{top}}(T, \mathcal{U}) = \lim_{\alpha \to 0} \limsup_{n \to \infty} \frac{1}{n} \log \left(\sup_{x \in X} N(\mathcal{U}^n|\Psi_n(\alpha, x)) \right). \tag{12}$$

Proof. Combining (10) and (11), we have

$$N_{\max}(n, \mathcal{U}) \le N(\mathcal{U}^n|\Psi_n(\alpha)) \le K \cdot N_{\max}(n, \mathcal{U})$$

for each fixed $0 < \alpha < \epsilon/4$ and all $n > 1$, where ϵ is a Lebesgue number of $\mathcal{U}$ and $K = K(\alpha)$ in Lemma 4.1 is independent of n. This implies that

$$\limsup_{n \to \infty} \frac{1}{n} \log N_{\max}(n, \mathcal{U}, \alpha) = \limsup_{n \to \infty} \frac{1}{n} \log N(\mathcal{U}^n|\Psi(\alpha)) \tag{13}$$

for all positive number $0 < \alpha < \epsilon/2$. Thus, we have

$$\begin{aligned} h_{\text{top}}(T, \mathcal{U}) &= \lim_{\alpha \to 0} \lim_{n \to \infty} \frac{1}{n} \log N(\mathcal{U}^n|\Psi_n(\alpha)) \quad (\text{by Theorem 2.3}) \\ &= \lim_{\alpha \to 0} \limsup_{n \to \infty} \frac{1}{n} \log N(\mathcal{U}^n|\Psi_n(\alpha)) \\ &= \lim_{\alpha \to 0} \limsup_{n \to \infty} \frac{1}{n} \log N_{\max}(n, \mathcal{U}, \alpha) \quad (\text{by } (4.6)) \\ &= \lim_{\alpha \to 0} \limsup_{n \to \infty} \frac{1}{n} \log \left(\sup_{x \in X} N(\mathcal{U}^n|\Psi_n(\alpha, x)) \right) \quad (\text{by } (4.1)) \end{aligned}$$

This completes the proof. □

Theorem 4.3. *Let (X, T) be a topological dynamical system. If $\mathcal{U} \in \mathcal{C}_X^o$, then we have*

$$h_{\text{top}}(T, \mathcal{U}) = \sup_{x \in X} \lim_{\alpha \to 0} \limsup_{n \to \infty} \frac{1}{n} \log N(\mathcal{U}^n | \Psi_n(\alpha, x)). \tag{14}$$

Proof. It follows directly from (10) and (12) that

$$\begin{aligned} h_{\text{top}}(T, \mathcal{U}) &= \lim_{\alpha \to 0} \limsup_{n \to \infty} \frac{1}{n} \log \left(\sup_{x \in X} N(\mathcal{U}^n | \Psi_n(\alpha, x)) \right) \\ &\geq \sup_{x \in X} \lim_{\alpha \to 0} \limsup_{n \to \infty} \frac{1}{n} \log N(\mathcal{U}^n | \Psi_n(\alpha, x)). \end{aligned}$$

Now we start to prove the converse inequality.

Note that for the given $n \geq 1$ and $\alpha > 0$, there is a point $y(n, \mathcal{U}, \alpha) \in X$ such that

$$N(\mathcal{U}^n | \Psi_n(\alpha, y(n, \mathcal{U}, \alpha)) = \max_{x \in X} N(\mathcal{U}^n | \Psi_n(\alpha, x)).$$

Taking a sequence of integers $n_i = n_i(\alpha) \to \infty$ such that

$$\limsup_{n \to \infty} \frac{1}{n} \log N(\mathcal{U}^n | \Psi_n(\alpha, y(n, \mathcal{U}, \alpha))) = \lim_{i \to \infty} \frac{1}{n_i} \log N(\mathcal{U}^{n_i} | \Psi_{n_i}(\alpha, y(n_i, \mathcal{U}, \alpha))).$$

By restricting to a subsequence, we can assume without loss of generality that the sequence $y_i(\alpha) = y(n_i, \mathcal{U}, \alpha)$ converses to a limit $q(\alpha)$.

Let ϵ be a Lebesgue number of $\mathcal{U}$. If $0 < \beta < \epsilon/4$ and $d(y_i(\alpha), q(\alpha)) < \beta$, then $\mathcal{V}$ is a subcover of $\mathcal{U}^n$ that covers $\Psi_n(\alpha, y_i(\alpha))$ whenever $\mathcal{V}$ is a subcover of $\mathcal{U}^n$ that covers $\Psi_n(\alpha + \beta, q(\alpha))$. This implies that

$$N(\mathcal{U}^n | \Psi_n(\alpha + \beta, q(\alpha))) \geq N(\mathcal{U}^n | \Psi_n(\alpha, y_i(\alpha))) \tag{15}$$

whenever $d(y_i(\alpha), q(\alpha)) < \beta$.

Now we choose a sequence $\alpha_j \to 0$ such that $q(\alpha_j)$ converges to some point $q \in X$. Similar to the proof as above we have

$$N(\mathcal{U}^n | \Psi_n(\alpha_j + 2\beta, q) \geq N(\mathcal{U}^n | \Psi_n(\alpha_j, q(\alpha_j))) \tag{16}$$

whenever $d(q(\alpha_j), q) < \beta$. Combining inequalities (15) and (16), one has

$$N(\mathcal{U}^n | \Psi_n(\alpha_j + 2\beta, q)) \geq N(\mathcal{U}^n | \Psi_n(\alpha_j, y_i(\alpha_j))) \tag{17}$$

whenever $d(y_i(\alpha_j), q(\alpha_j)) < \beta$ and $d(q(\alpha_j), q) < \beta$. If j is a fixed integer with $d(q(\alpha_j), q) < \beta$, then (17) holds for all sufficiently large integers i. Thus,

$$\begin{aligned} &\limsup_{n \to \infty} \frac{1}{n} \log N(\mathcal{U}^n | \Psi_n(\alpha_j + 2\beta, q)) \\ &\qquad \geq \limsup_{i \to \infty} \frac{1}{n_i} \log N(\mathcal{U}^{n_i} | \Psi_{n_i}(\alpha_j + 2\beta, q)) \\ &\qquad \geq \lim_{i \to \infty} \frac{1}{n_i} \log N(\mathcal{U}^{n_i} | \Psi_{n_i}(\alpha_j, y_i(\alpha_j))) \\ &\qquad = \limsup_{n \to \infty} \frac{1}{n} \log \left(\max_{x \in X} N(\mathcal{U}^n | \Psi_n(\alpha_j), x)) \right). \end{aligned} \tag{18}$$

Now let $j \to \infty$ and use the fact that both sides (18) are nonincreasing as α decreases to conclude that

$$\begin{aligned}
&\limsup_{n\to\infty}\frac{1}{n}\log N(\mathcal{U}^n|\Psi_n(3\beta,q)) \\
&\qquad \geq \lim_{j\to 0}\limsup_{n\to\infty}\frac{1}{n}\log N\big(\mathcal{U}^n|\Psi_n\big(\alpha_j+2\beta,q\big)\big) \\
&\qquad \geq \lim_{j\to\infty}\limsup_{n\to\infty}\frac{1}{n}\log\left(\max_{x\in X}N(\mathcal{U}^n|\Psi_n\big(\alpha_j\big),x))\right) \\
&\qquad = \inf_{\alpha>0}\limsup_{n\to\infty}\frac{1}{n}\log\left(\max_{x\in X}N(\mathcal{U}^n|\Psi_n(\alpha,x))\right) \\
&\qquad = \lim_{\alpha\to 0}\limsup_{n\to\infty}\frac{1}{n}\log\left(\max_{x\in X}N(\mathcal{U}^n|\Psi_n(\alpha),x))\right).
\end{aligned} \tag{19}$$

Therefore, combining (12) and (19), we have

$$\begin{aligned}
h_{\text{top}}(T,\mathcal{U}) &= \lim_{\alpha\to 0}\text{limsup}_{n\to\infty}\frac{1}{n}\log\left(\max_{x\in X}N(\mathcal{U}^n|\Psi_n(\alpha),x))\right) \\
&\leq \inf_{\beta>0}\text{limsup}_{n\to\infty}\frac{1}{n}\log N(\mathcal{U}^n|\Psi_n(3\beta,q)) \\
&= \inf_{\alpha>0}\text{limsup}_{n\to\infty}\frac{1}{n}\log N(\mathcal{U}^n|\Psi_n(\alpha,q)) \\
&= \lim_{\alpha\to 0}\text{limsup}_{n\to\infty}\frac{1}{n}\log N(\mathcal{U}^n|\Psi_n(\alpha,q)) \\
&\leq \sup_{x\in X}\lim_{\alpha\to 0}\text{limsup}_{n\to\infty}\frac{1}{n}\log N(\mathcal{U}^n|\Psi_n(\alpha,x)).
\end{aligned} \tag{20}$$

This completes the proof. □

5. Partial entropy and bundle-like entropy for nonautonomous discrete dynamical systems

In [38, 41], topological entropy for certain nonautonomous discrete dynamical system was defined and studied. In this section, we study the topological entropy for nonautonomous discrete dynamical systems by introducing two entropy-like invariants called the partial entropy and bundle-like entropy as being motivated by the idea of [1, 39].

5.1 Topological entropy for nonautonomous discrete dynamical systems

Let $\mathbf{X}$ be a collection of countable infinitely many compact metric space (X_i, d_i) and $\mathbf{F}$ be a collection of countable infinite many continuous maps $f_i : X_i \to X_{i+1}$, $i = 1, 2, \cdots$. Then the pair $(\mathbf{X}, \mathbf{F})$ is called a *nonautonomous discrete dynamical system*.

For any integer $n \geq 1$, we define a metric $\tilde{d}_n$ on $\prod_{i=1}^n X_i$ as follows: for any two points $\tilde{x}_n = (x_1, x_2, \cdots, x_n), \tilde{y}_n = (y_1, y_2, \cdots, y_n) \in \prod_{i=1}^n X_i$,

$$\tilde{d}_n(\tilde{x}_n, \tilde{y}_n) = \max_{1 \le i \le n} d_i(x_i, y_i).$$

Fixing an integer $n \geq 1$ and a positive number ϵ. A subset Z of $\prod_{i=1}^{n} X_i$ is called $\tilde{d}_n$-(n, ϵ)-*separated* if for any two distinct points $\tilde{x}_n, \tilde{y}_n \in Z$ we have $\tilde{d}_n(\tilde{x}_n, \tilde{y}_n) > \epsilon$. Denote the maximal cardinality of any $\tilde{d}_n$-separated subset of Z by $s(n, \epsilon, Z)$. A subset $W \subset Z$ is called $\tilde{d}_n$-(n, ϵ)-*spanning for* Z if for each $\tilde{z}_n \in Z$, there is a $\tilde{w}_n \in W$ such that $\tilde{d}_n(\tilde{z}_n, \tilde{w}_n) < \epsilon$. Denote the minimal cardinality of any $\tilde{d}_n$-spanning subset of Z by $r(n, \epsilon, Z)$.

The following result is trivial, so we omit its detail proof.

Lemma 5.1. *Suppose that n is a positive integer and Z is a nonempty subset of* $\prod_{i=1}^{n} X_i$. *Then for each* $\epsilon > 0$, *we have*

$$r(n, \epsilon, Z) \leq s(n, \epsilon, Z) \leq r(n, \epsilon/2, Z).$$

For each $n \geq 1$ let Z_n be a nonempty subset of $\prod_{i=1}^{n} X_i$. Then it follows immediately from Lemma 5.1 that

$$\lim_{\epsilon \to 0} \limsup_{n \to \infty} \frac{1}{n} \log r(n, \epsilon, Z_n) = \lim_{\epsilon \to 0} \limsup_{n \to \infty} \frac{1}{n} \log s(n, \epsilon, Z_n). \tag{21}$$

Given a nonautonomous discrete dynamical system $(\mathbf{X}, \mathbf{F})$, denoted by $O_{n, \mathbf{F}}$ or O_n for short the set of all orbit segments of length n for each $n \geq 1$, i.e.,

$$O_n = O_{n, \mathbf{F}} := \{(x_1, x_2, \cdots, x_n) : x_1 \in X_1 \ \ and \ \ x_{i+1} = f_i(x_i), i = 1, 2, \cdots, n-1\}.$$

Then the common limit in (21) by taking $Z_n = O_n$ is defined to be the *topological entropy of* $(\mathbf{X}, \mathbf{F})$, written $h_{\text{top}}(\mathbf{X}, \mathbf{F})$ or $h_{\text{top}}(\mathbf{F})$ for short if there is no confusion.

5.2 Partial entropy and bundle-like entropy

Let $(\mathbf{X}, \mathbf{F})$ be a nonautonomous discrete dynamical system. A collection $\mathcal{P} = (\mathcal{P}_i : i \geq 1)$ is said to be a *cover* of $\mathbf{X}$ if each $\mathcal{P}_i$ covers X_i, respectively. We now define two entropies, partial entropy and bundle-like entropy, for $(\mathbf{X}, \mathbf{F})$ relative to $\mathcal{P}$.

For any integer $n \geq 1$ and $D \in \mathcal{P}_n$, let $W_n(D) \subset \prod_{i=1}^{n} X_i$ denote the set of all orbit segments of length that end at some point $x_n \in D$, i.e.,

$$W_n(D) = \{(x_1, x_2, \cdots, x_n) \in O_n : x_n \in D\}.$$

Put $s_{\max, \mathcal{P}_n}(n, \epsilon) = \sup_{D \in \mathcal{P}_n} s(n, \epsilon, W_n(D))$. Define the entropy by

$$h_{p, \mathcal{P}}(\mathbf{X}, \mathbf{F}) = \lim_{\epsilon \to 0} \limsup_{n \to \infty} \frac{1}{n} \log s_{\max, \mathcal{P}_n}(n, \epsilon),$$

which is called the *partial entropy of* $(\mathbf{X}, \mathbf{F})$ *relative to* $\mathcal{P}$ and written shortly by $h_{p, \mathcal{P}}(\mathbf{F})$ if there is no confusion.

Let $O_{n, \mathcal{P}_n} = \{W_n(D) : D \in \mathcal{P}_n\}$. For any two elements, $W_n(D)$ and $W_n(E)$ of $O_{n, \mathcal{P}_n}$, denoted by $d_H(W_n(D), W_n(E))$, the usual Hausdorff metric between them is based upon metric $\tilde{d}_n$ of $\prod_{i=1}^{n} X_i$ defined as before and by $s(n, \epsilon, O_{n, \mathcal{P}_n})$ the maximum cardinality of any d_H-(n, ϵ)-separated subset of $O_{n, \mathcal{P}_n}$. Define the entropy by

$$h_{b,\mathcal{P}}(\mathbf{X},\mathbf{F}) = \lim_{\epsilon\to 0} \limsup_{n\to\infty} \frac{1}{n} \log s\left(n, \epsilon, O_{n,\mathcal{P}_n}\right),$$

which is called the *bundle-like entropy of* $(\mathbf{X},\mathbf{F})$ *relative to* $\mathcal{P}$ and written shortly by $h_{b,\mathcal{P}}(\mathbf{F})$ if there is no confusion.

Also, we have the spanning set versions of definitions of $h_{p,\mathcal{P}}(\mathbf{F})$ and $h_{b,\mathcal{P}}(\mathbf{F})$, respectively.

5.3 Some relationships between $h_{\text{top}}(\mathbf{F})$ and $h_{p,\mathcal{P}}(\mathbf{F})$

Theorem 5.2. *Let* $(\mathbf{X},\mathbf{F})$ *be a nonautonomous discrete dynamical system, and* $\mathcal{P} = (\mathcal{P}_i : i \geq 1)$ *be a cover of* $\mathbf{X}$. *Then we have*

$$h_{p,\mathcal{P}}(\mathbf{F}) \leq h_{\text{top}}(\mathbf{F}) \leq h_{b,\mathcal{P}}(\mathbf{F}) + h_{p,\mathcal{P}}(\mathbf{F}).$$

Proof. Note that $s_{\max,\mathcal{P}_n}(n,\epsilon) \leq s(n,\epsilon,O_n)$ for any cover $\mathcal{P}$ of $\mathbf{X}$ and any $\epsilon > 0$. Then the former inequality is obtained. Now we show the later one. If $h_{b,\mathcal{P}}(\mathbf{F}) = \infty$, then there is nothing to prove. Now assuming $h_{b,\mathcal{P}}(\mathbf{F}) < \infty$.

Fixing a sufficiently small $\epsilon > 0$ and an integer $n \geq 1$, let Y be a d_H-(n,ϵ)-separated subset of $O_{n,\mathcal{P}_n}$ with cardinality $s\left(n,\epsilon,O_{n,\mathcal{P}_n}\right)$. For each $W_n(D) \in Y$, let $M(D)$ be a $\tilde{d}_n$-(n,ϵ)-separated subset of $W_n(D)$ with cardinality $s(n,\epsilon,W_n(D))$. Put $M = \cup_{W_n(D)\in Y} M(D)$. We claim that M is a $\tilde{d}_n$-$(n,3\epsilon)$-spanning subset of O_n.

In fact, for any $x = (x_1, x_2, \cdots, x_n) \in O_n$, since Y is a d_H-(n,ϵ)-separated subset of $O_{n,\mathcal{P}_n}$ with maximum cardinality and $\mathcal{P}_n$ covers X_n, there is an $E \in \mathcal{P}_n$ with $x_n \in E$ and a $W_n(D) \in Y$ such that $d_H(W_n(D), W_n(E)) \leq \epsilon$. Then it follows that there is a $y = \left(y_1, y_2, \cdots, y_n\right) \in W_n(D)$ such that $\tilde{d}_n(x,y) \leq \epsilon$. Also note that $M(D)$ is a $\tilde{d}_n$-(n,ϵ)-separated subset of $W_n(D)$ with maximum cardinality; there is a $z \in M(D)$ such that $\tilde{d}_n(y,z) \leq \epsilon$. Hence we have

$$\tilde{d}_n(x,z) \leq \tilde{d}_n(x,y) + \tilde{d}_n(y,z) < 3\epsilon.$$

This yields the claim that M is a $\tilde{d}_n$-(n,ϵ)-spanning subset of O_n. So we have $r(n,3\epsilon,O_n) \leq |M|$, where M denotes the cardinality of M. Using the claim we have

$$\begin{aligned} r(n,3\epsilon,O_n) &\leq |M| \leq |Y| \cdot \max\{|M(D)| : W_n(D) \in Y\} \\ &\leq s\left(n,\epsilon,O_{n,\mathcal{P}_n}\right) \cdot s_{\max,\mathcal{P}_n}(n,\epsilon). \end{aligned}$$

Taking limits as the requirements of the related definitions of entropies establishes the desired inequality. This completes the proof. □

Let $\mathcal{P}(\delta)$ be a finite cover of a compact metric space X consisting of open balls with radius less than some $\delta > 0$. Write $\mathbf{F}_X = \{f_i : f_i : X \to X \text{ is continous}, i \geq 1\}$ and $\mathcal{P}_X(\delta) = (\mathcal{P}(\delta), \mathcal{P}(\delta), \cdots)$.

Theorem 5.3.

$$h_{\text{top}}(\mathbf{F}_X) = h_{p,\mathcal{P}_X(\delta)}(\mathbf{F}_X) = \lim_{\epsilon\to 0} \lim_{\delta\to 0} \limsup_{n\to\infty} \frac{1}{n} \log s_{\max,\mathcal{P}(\delta)}(n,\epsilon).$$

Proof. Note that $\lim_{n\to\infty} \frac{1}{n} \log|\mathcal{P}(\delta)| = 0$. Then, by Theorem 5.2, we have the former equality. Now we show the later equality.

Clearly, $s(n,\epsilon,O_n) \geq s_{\max,\mathcal{P}_X}(\delta)(n,\epsilon)$ for any $\delta > 0$, so we have

$$\limsup_{n\to\infty} \log s(n,\epsilon,O_n) \geq \lim_{\delta\to 0}\limsup_{n\to\infty}\frac{1}{n}\log s_{\max,\,\mathcal{P}_X(\delta)}(n,\epsilon).$$

This implies

$$h_{\text{top}}(\mathbf{F}_X) \geq \lim_{\varepsilon\to 0}\lim_{\delta\to 0}\limsup_{n\to\infty}\frac{1}{n}\log s_{\max,\,\mathcal{P}_X(\delta)}(n,\varepsilon). \tag{22}$$

On the other hand, from the proof of Theorem 5.2, it follows that

$$r(n,3\epsilon,O_n) \leq s\big(n,\epsilon,O_{n,\,\mathcal{P}(\delta)}\big)\cdot s_{\max,\,\mathcal{P}_X(\delta)}(n,\epsilon)$$

for any integer $n\geq 1$, any sufficiently small $\epsilon>0$ and any $\delta>0$. Noting that $s\big(n,\epsilon,O_{n,\,\mathcal{P}(\delta)}\big)\leq|\mathcal{P}(\delta)|$ for any integer $n\geq 1$, then we have

$$\limsup_{n\to\infty}\frac{1}{n}\log r(n,3\epsilon,O_n) \leq \lim_{\delta\to 0}\limsup_{n\to\infty}\frac{1}{n}\log s_{\max,\,\mathcal{P}_X(\delta)}(n,\epsilon).$$

This implies

$$h_{\text{top}}(\mathbf{F}_X) \leq \lim_{\epsilon\to 0}\lim_{\delta\to 0}\limsup_{n\to\infty}\frac{1}{n}\log s_{\max,\,\mathcal{P}_X(\delta)}(n,\epsilon). \tag{23}$$

Thus, combining (22) and (23) gets the later equality. This completes the proof. □

Remark 5.4. The first equality of Theorem 5.3 is in fact a simpler version of Theorem 7.6 of [40] (a useful result for calculating the classical topological entropy) when restricting to the autonomous discrete dynamical systems.

Given a nonautonomous discrete dynamical system $(\mathbf{X},\mathscr{F})$, when does $h_{\text{top}}(\mathbf{F}) = h_{p,\,\mathcal{P}}(\mathbf{F})$ for any cover $\mathcal{P}$ of $\mathbf{X}$? The following theorem gives an answer to this question.

Theorem 5.5. *Let* $(\mathbf{X},\mathbf{F})$ *be a nonautonomous discrete dynamical system. Then* $h_{\text{top}}(\mathbf{F}) = h_{p,\,\mathcal{P}}(\mathbf{F})$ *for any cover* $\mathcal{P}$ *of* $\mathbf{X}$ *if the following conditions hold*:

(1) *For each integer* $i\geq 1$, *there exists* $\delta_i>0$ *such that* $d_{i+1}\big(f_i(x),f_i(y)\big)\geq d_i(x,y)$ *whenever* $d_i(x,y)\leq\delta_i$ *for* $x,y\in X_i$.

(2) *For each integer* $i\geq 1$, *every* $x\in X_{i+1}$ *has an open neighborhood* U_x *whose preimage* $f_i^{-1}(U_x)$ *is an union of disjoint open sets on each of which* f_i *is a homeomorphism.*

(3) $\limsup_{n\to\infty}\frac{1}{n}\log N(\epsilon_n,X_n)=0$ *for every monotonic decreasing sequence* $\{\epsilon_n\}$ *with* $\lim_{n\to\infty}\epsilon=0$, *where each* $N(\epsilon_n,X_n)$ *denotes the minimal cardinality of the open cover of X consisting of open* ϵ_n*-ball for the compact metric space* X_n.

Proof. It suffices to show that $h_{b,\,\mathcal{P}}(\mathbf{F})=0$ for any cover $\mathcal{P}$ of $\mathbf{X}$ by Theorem 5.2. Let $\mathcal{P}_{\max}=\{\mathcal{P}_{i,\,\max}: i\geq 1\}$ be the cover of $\mathbf{X}$ in which each $\mathcal{P}_{i,\,\max}$ cover X_i consisting to singletons of X_i, i.e., $\mathcal{P}_{i,\,\max}=\{\{z\}: z\in X_i\}$. It is easy to see that $h_{b,\,\mathcal{P}}(\mathbf{F})\leq h_{b,\,\mathcal{P}_{\max}}(\mathbf{F})$ for any cover $\mathcal{P}$ of $\mathbf{X}$. So from Theorem 5.2, it follows that what we want to prove is $h_{b,\,\mathcal{P}_{\max}}(\mathbf{F})=0$.

For each $n\geq 2$, by condition (1), there exists a $\delta_{n-1}>0$ such that

$$d_n\big(f_{n-1}(x),f_{n-1}(y)\big)\geq d_{n-1}(x,y)$$

for any $x,y\in X_{n-1}$ whenever $d_{n-1}(x,y)\leq\delta_{n-1}$. Also, by condition (2) and the compactness of X_n, there exists an $\epsilon_n>0$ such that the ϵ_n-ball $B(x_n,\epsilon_n)$ about any point $x_n\in X_n$ has preimage $f_{n-1}^{-1}(B(x_n,\epsilon))$ equals the union of disjoint open sets of diameter less than δ_{n-1}. Then we get a sequence $\{\epsilon_n\}$. Furthermore, we can take ϵ_n such that $\{\epsilon_n\}$ is monotonic decreasing sequence and $\lim_{n\to\infty}\epsilon_n = 0$.

Now, given $y_n \in X_n$ and $\tilde{x} = (x_1, x_2, \cdots, x_n) \in W_n(B(y_n, \epsilon_n,))$, we want to find a point $\tilde{y} = (y_1, y_2, \cdots, y_n) \in O_n$ with $\tilde{d}_n(\tilde{x}, \tilde{y}) = d_n(x_n, y_n)$ and then $\tilde{d}_n(\tilde{x}, \tilde{y}) < \epsilon_n$. In fact, for $1 < k < n$, we can easily find a point $(y_k, \cdots, y_n) \in \prod_{i=k}^n X_i$ with $d_j(x_j, y_j) \le \epsilon_j$ and $d_{j+1}(x_{j+1}, y_{j+1}) \ge d_j(x_j, y_j)$, for $j = n-1, n-2, \cdots, k$. Let V be the piece of $f_{k-1}^{-1}(B(x_k, \epsilon_k))$ with $x_{k-1} \in V$. Since $y_k \in B(x_k, \epsilon_k)$, there is a unique point $y_{k-1} \in V \cap f_{k-1}^{-1}(y_k)$ such that $d_{k-1}(x_{k-1}, y_{k-1}) < \delta_{k-1}$. Then we have

$$d_{k-1}(x_{k-1}, y_{k-1}) \le d_k(x_k, y_k) \le d_n(x_n, y_n) < \epsilon_n < \epsilon_{k-1}.$$

This argument shows that $r(n, \epsilon_n, O_{n, comma\, \max}) \le N(\epsilon_n, X_n)$. Thus, by condition (3), we get

$$\limsup_{n\to\infty} \frac{1}{n} \log r(n, \epsilon_n, O_{n, comma\, \max}) \le \limsup_{n\to\infty} \frac{1}{n} \log N(\epsilon_n, X_n) = 0.$$

For any sufficiently small $\epsilon > 0$, there exists $N > 0$ such that $\epsilon_n < \epsilon$ for any $n \ge N$. Then we have $r\left(n, \epsilon, O_{n, \mathcal{P}_{n},\, \max}\right) \le r\left(n, \epsilon_n, O_{n, \mathcal{P}_{n},\, \max}\right)$ and hence $h_{b, \mathcal{P}_{\max}}(\mathbf{F}) = 0$. This completes the proof. □

6. Conclusion

Several important entropy-like invariants based on the preimage structure for non-invertible maps have been defined and studied by some authors. In this chapter, we first further study the preimage entropy for topological dynamical system from the view of localization. We show that the topological entropy for an open cover can be characterized by pseudo-orbits (Theorems 2.3, 4.2, and 4.3). We also establish an inequality relating the topological entropy for open covers and several preimage entropy invariants, which is viewed as the local version of the Hurley's inequality (Theorem 3.1). Finally, we discuss the topological entropy for nonautonomous discrete dynamical systems by introducing two entropy-like invariants called the partial entropy and bundle-like entropy. We establish some relationships among such two invariants and the topological entropy (Theorem 5.2, 5.3, and 5.5).

Acknowledgements

This work was carried out when Kesong Yan visited the Michigan State University. Kesong Yan sincerely appreciates the warm hospitality of Professor Huyi Hu. We thank the anonymous referees for their useful comments and helpful suggestions that improved the manuscript. The authors are supported by NNSF of China (11861010,11761012) and NSF for Distinguished Young Scholar of Guangxi Province (2018GXNSFFA281008). The first author is supported by the Cultivation Plan of Thousands of Young Backbone Teachers in Higher Education Institutions of Guangxi Province, Program for Innovative Team of Guangxi University of Finance and Economics, and Project of Guangxi Key Laboratory Cultivation Base of Cross-border E-commerce Intelligent Information Processing (201801ZZ03).

Author details

Kesong Yan and Fanping Zeng*
School of Information and Statistics, Guangxi University of Finance and Economics, Nanning, Guangxi, PR China

*Address all correspondence to: fpzeng@gxu.edu.cn

IntechOpen

References

[1] Hurley H. On topological entropy of maps. Ergodic Theory and Dynamical Systems. 1995;**15**:557-568. DOI: 10.1017/S014338570000852X

[2] Adler RL, Konheim AG, McAndrew MH. Topological entropy. Transactions of the American Mathematical Society. 1965;**114**: 309-319. DOI: 10.1090/S0002-9947-1965-0175106-9

[3] Dinaburg EI. On the relations among various entropy characteristic of dynamical systems. Izvestiya Rossiiskoi Akademii Nauk. Seriya Matematicheskaya. 1971;**35**:324-366

[4] Bowen R. Entropy for group endomorphisms and homogeneous spaces. Transactions of the American Mathematical Society. 1971;**153**:401-414. DOI: 10.1090/S0002-9947-1971-0274707-X

[5] Langevin R, Walczak P. Entropie d'une dynamique [Entropy of a dynamic]. Comptes Rendus de I' Académie des Sciences-Series I-Mathematics. 1991;**312**: 141-144. (in French)

[6] Ghys E, Langevin R, Walczak P. Entroie geoetrique des feuilletages [Geometric entropy of foliations]. Acta Math. 1988;**160**:105-142 (in French). DOI: 10.1007/BF02392274

[7] Langevin R, Przytycki F. Entropie de I'image inverse d'une application. [Entropy of the inverse of a mapping]. Bulletin de la Société Mathématique de France. 1992;**120**:237-250. DOI: 10.24033/bsmf.2185 (in French)

[8] Nitecki Z, Przytycki F. Preimage entropy for mappings. International Journal of Bifurcation and Chaos. 1999; **9**:1815-1843. DOI: 10.1142/S0218127499001309

[9] Biś A. Entropies of a semigroup of maps. Discrete and Continuous Dynamical Systems. 2004;**11**:639-648. DOI: 10.3934/dcds.2004.11.639

[10] Nitecki Z. Topological entropy and the preimage structure of maps. Real Analysis Exchange. 2003;**29**:9-43. DOI: 10.14321/realanalexch.29.1.0009

[11] Yan K, Zeng F. Variational principles of partial pre-image entropy and conditional pre-image entropy. Scientia Sinica Mathematica. 2019;**49**: 681-698. DOI: 10.1360/SCM-2016-0381. (in Chinese)

[12] Cheng W. Relations among conditional entropy, topological entropy and pointwise preimage entropy [thesis]. East Lansing: Michigan State University; 2004

[13] Cheng W. Pre-image entropy. Ergodic Theory and Dynamical Systems. 2005;**25**:1091-1113. DOI: 10.1017/S0143385704000240

[14] Cheng W. Forward generator for preimage entropy. Pacific Journal of Mathematics. 2006;**223**:5-16. DOI: 10.2140/pjm.2006.223.5

[15] Cheng W. Two-point pre-image entropy. Discrete and Continuous Dynamical Systems. 2007;**17**:107-119. DOI: 10.3934/dcds.2007.17.107

[16] Cheng W. On preimage entropy of a sequential mapping. Open Systems and Information Dynamics. 2008;**15**:345-357. DOI: 10.1142/S1230161208000237

[17] Cheng W. Pre-image entropy for free semigroup actions. Chaos, Solitons and Fractals. 2016;**91**:286-290. DOI: 10.1016/j.chaos.2016.06.011

[18] Fiebig D, Fiebig U, Nitecki Z. Entropy and preimage sets. Ergodic Theory and Dynamical Systems. 2003; **23**:1785-1806. DOI: 10.1017/S0143385703000221

[19] Yan K, Zeng F, Zhang G. Equilibrium states for pre-image pressure. Nonlinear Dynamics and Systems Theory. 2010;**10**:409-416

[20] Yan K, Zeng F, Zhang G. Differentiability properties of the pre-image pressure. Discrete Dynamics in Nature and Society; **2012**. DOI: 10.1155/2012/951691, Article ID 951691, 14 pages

[21] Zeng F, Yan K, Zhang G. Pre-image pressure and invariant measures. Ergodic Theory and Dynamical Systems. 2007;**27**:1037-1052. DOI: 10.1017/S0143385706000812

[22] Zhu Y. Preimage entropy for random dynamical systems. Discrete and Continuous Dynamical Systems. 2007;**18**:829-851. DOI: 10.3934/dcds.2007.18.829

[23] Zhu Y, Li Z, Li X. Preimage pressure for random transformations. Ergodic Theory and Dynamical Systems. 2009; **29**:1669-1687. DOI: 10.1017/S0143385708000758

[24] Blanchard F. Fully positive topological entropy and topological mixing. In: Walters P, editor. Symbolic Dynamics and its Applications. Contemporary Mathematics. Vol. 135. Providence: RI: American Mathematical Society; 1992. pp. 95-105. DOI: 10.1090/conm/135

[25] Blanchard F, Lacroix Y. Zero-entropy factors of topological flows. Proceedings of the American Mathematical Society. 1993;**119**:985-992. DOI: 10.1090/S0002-9939-1993-1155593-2

[26] Blanchard F. A disjointness theorem involving topological entropy. Bulletin de la Société Mathématique de France. 1993;**121**:565-578. DOI: 10.24033/bsmf.2216

[27] Blanchard F, Host B, Maass A, Martinez S, Rudolph D. Entropy pairs for a measure. Ergodic Theory and Dynamical Systems. 1995;**15**:621-632. DOI: 10.1017/S0143385700008579

[28] Glasner E, Weiss B. Topological entropy of extensions. In: Petersen Karl E, editor. Ergodic Theory and its Connections with Harmonic Analysis. Cambridge: Cambridge University Press; 1995. pp. 299-307. DOI: 10.1017/CBO9780511574818.011

[29] Huang W, Ye X. A local variational relation and applications. Israel Journal of Mathematics. 2006;**151**:237-279. DOI: 10.1007/BF02777364

[30] Glasner E, Ye X. Local entropy theory. Ergodic Theory and Dynamical Systems. 2009;**29**:321-356. DOI: 10.1017/S0143385708080309

[31] Bowen R. Equilibrium States and the Ergodic Theory of Axiom Diffeomorphisms, Lecture Notes in Math. Second revised ed. Vol. 470. New York: Springer-Verlag, Berlin; 2008. 75 p. DOI: 10.1007/BFb0081284

[32] Conley C. Isolated Invariant Sets and the Morse Index. Vol. 38. Providence: CBMS Regional Conference Series in Mathematics, American Mathematical Society; 1978. 89 p. DOI: 10.1090/cbms/038

[33] Hammel SM, Yorke JA, Grebogi C. Do numerical orbits of chaotic dynamical processes represent true orbits? Journal of Complexity. 1987;**3**:136-145. DOI: 10.1016/0885-064X(87)90024-0

[34] Hammel SM, Yorke JA, Grebogi C. Numerical orbits of chaotic processes represent true orbits. Bulletin of the American Mathematical Society. 1988; **19**:465-469

[35] Barge M, Swanson R. Rotation shadowing properties of circle and annulus maps. Ergodic Theory and Dynamical Systems. 1988;**8**:509-521. DOI: 10.1017/S0143385700004661

[36] Misiurewicz M. Remark on the definition of topological entropy. In: Lara-Carrero L, Lewowicz J, editors. Dynamical Systems and Partial Differential Equations (Caracas, 1984). Caracas: Equinoccio; 1986. pp. 65-67

[37] Barge M, Swanson R. Pseudo-orbits and topological entropy. Proceedings of the American Mathematical Society. 1990;**109**:559-566. DOI: 10.1090/S0002-9939-1990-1012923-7

[38] Kolyada S, Snoha L. Topological entropy of nonautonomous dynamical systems. Random & Computational Dynamics. 1996;**4**:205-233

[39] Zeng F. A note on topological entropy of maps. Northeastern Mathematical Journal. 1997;**13**:250-255. DOI: 10.13447/j.1674-5647.1997.04.012

[40] Walters P. An introduction to ergodic theory. In: Graduate Texts in Mathematics. Vol. 79. New York: Berlin: Springer-Verlag; 1982. 256 p. DOI: 10.1007/978-1-4612-5775-2

[41] Kolyada S, Snoha L. On topological dynamics of sequences of continuous map. International Journal of Bifurcation and Chaos. 1995;**5**: 1437-1438. DOI: 10.1142/S0218127495001125

Chapter 3

Determination of Stresses in Composite Plates with Holes and Cracks Based on Singular Integral Equations

Olesia Maksymovych and Adam Podhorecki

Abstract

The problems of determination of stresses at crack in bounded plates with holes of different shapes under the action of concentrated forces or distributed forces at its boundary are considered. The study is performed by the singular integral equation method. They were determined based on the established interdependences between the Lekhnitskii potentials and the stress and strain. The numerical method for solving integral equations is developed based on the quadrature method for the systems of holes and cracks. The eigen solutions of the problem were taken into account in this method. The research of stresses at cracks in samples which are used in experimental studies of crack fracture resistance was performed.

Keywords: stress intensity factors (SIF), composite plates, holes, cracks, crack fracture resistance, BIEM, stress-strain state (SSS)

1. Introduction

The boundary integral equation method (BIEM) is widely used to study the stress-strain state (SSS) of anisotropic plates with holes [1–3] and cracks [4–9]. The integral equations for anisotropic plates are usually determined based on the Somigliana identity. Such equations for plates with given stresses at the boundaries of the plate are hypersingular. At the same time, the same problem for isotropic plates is reduced to singular integral equations [10, 11], for which simple numerical algorithms for solving with given precision are obtained.

In [12, 13], the simple dependencies between the Lekhnitskii complex potentials and stress and strain are obtained. In a simple form based on them and the Cauchy theorem, the integral equations are written for anisotropic plates with holes [12, 13] and cracks [14–16]. We will use the established dependencies for the construction and regularization of integral equations for anisotropic plates with holes and cracks.

For conducting experimental studies of crack fracture resistance on experimental samples in relation to isotropic materials, theoretical estimates for stresses at cracks are performed.

For such materials, the stresses in samples of different shapes with cracks under the action of stretching or compressing concentrated forces are studied in detail [10]. The experimental samples for the experimental determination of the characteristics of crack fracture resistance of various types of materials are made based on performed studies. We perform similar studies for composite samples.

2. The integral representations for anisotropic plates with holes and cracks

We consider a plate which is weakened with a system hole with boundaries $L_1, ..., L_J$ $(j = 1, ..., J)$, and cracks are placed along curves Γ_k $(k = 1, ..., K)$. The L_0 is the outer boundary of plates. Assume (**Figure 1**) that a plate is loaded with concentrated forces $(X_j, Y_j), j = 1, ..., M$ acting at the points (a_j, b_j); tractions (X_T, Y_T) are applied to the crack edges, which are accepted the same on its opposite edges; and tractions (X_L, Y_L) are applied to the boundaries of the holes and plate.

2.1 Governing equations

Let us start from the Lekhnitskii complex potentials $\Phi(z_1), \Psi(z_2)$, where $z_j = x + s_j y$ and $s_j, j = 1, 2$ are roots with positive imaginary part of the characteristic equation $\Delta(s) = 0$ [10]:

where

and

$$\Delta(s) = \alpha_{11}s^4 - 2\alpha_{16}s^3 + (2\alpha_{12} + \alpha_{66})s^2 - 2\alpha_{26}s + \alpha_{22} \tag{1}$$

α_{ij} are elastic compliances which are included in the Hooke's law [10]:

$$\varepsilon_x = a_{11}\sigma_x + a_{12}\sigma_y + a_{16}\tau_{xy}, \varepsilon_y = a_{12}\sigma_x + a_{22}\sigma_y + a_{26}\tau_{xy}, \gamma_{xy} = a_{16}\sigma_x + a_{26}\sigma_y + a_{66}\tau_{xy},$$

where $\varepsilon_x, \varepsilon_y, \gamma_{xy}$ are strains and $\sigma_x, \sigma_y, \tau_{xy}$ are stresses.

Consider an arbitrary path Γ, which belongs to the domain D occupied by the plate, and select a positive direction of traversal (**Figure 2**).

Then introduce in consideration the stress vectors $\vec{q_\Gamma}$ at the plane tangent to the curve. The normal to it is located right relative to the selected direction of traversal. The projections (X_Γ, Y_Γ) of stress vectors $\vec{q_\Gamma}$ and derivatives of displacements (u, v) with respect to an arc coordinate at the curve through Lekhnitskii complex potentials are determined by the formula [17]:

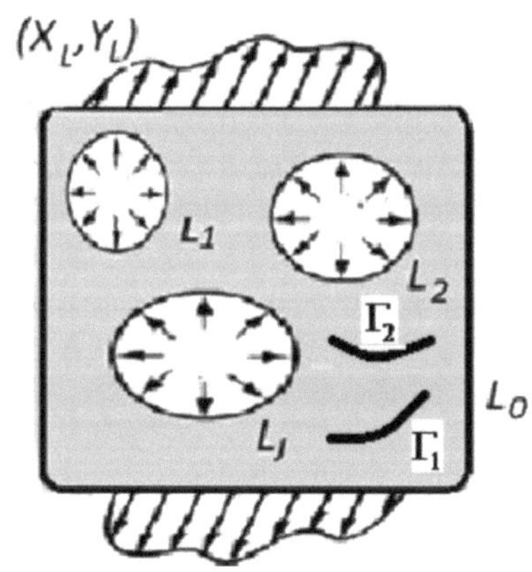

Figure 1.
Scheme of the problem.

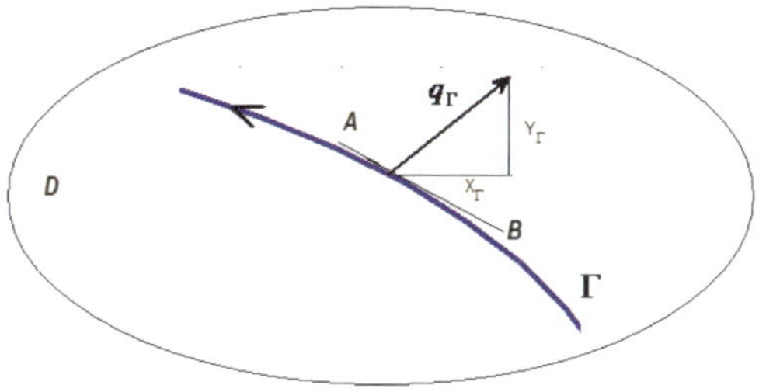

Figure 2.
$\vec{q_\Gamma}$ is the stress vector at plane AB.

$$Y_\Gamma = -2\mathrm{Re}\left[\Phi(z_1)z_1' + \Psi(z_2)z_2'\right],\;\; X_\Gamma = 2\mathrm{Re}\left[s_1\Phi(z_1)z_1' + s_2\Psi(z_2)z_2'\right], \tag{2}$$

$$u' = 2\mathrm{Re}\left[p_1\Phi(z_1)z_1' + p_2\Psi(z_2)z_2'\right],\;\; v' = 2\mathrm{Re}\left[q_1\Phi(z_1)z_1' + q_2\Psi(z_2)z_2'\right], \tag{3}$$

where $u' = du/ds$, $v' = dv/ds$ and $z_j' = dx/ds + s_j dy/ds$, where ds is a differential of arc at Γ.

The stress vectors $q_\Gamma(z) = X_\Gamma + iY_\Gamma$ at path Γ are determined using the formulas (2) by the formula:

$$q_\Gamma = (s_1 - i)z_1'\Phi(z_1) + (\overline{s_1} - i)\overline{z_1'}\,\overline{\Phi(z_1)} + (s_2 - i)z_2'\Psi(z_2) + (\overline{s_2} - i)\overline{z_2'\Psi(z_2)}. \tag{4}$$

Assume that the vectors (X, Y) and (u, v) are known at path Γ. Then based on Eqs. (2) and (3) at Γ one has [12, 15]

$$\Phi(z_1) = \frac{-v' + s_1u' + p_1X + q_1Y}{\Delta_1 z_1'},\;\; \Psi(z_2) = \frac{-v' + s_2u' + p_2X + q_2Y}{\Delta_2 z_2'} \tag{5}$$

where $\Delta_j = \Delta'(s_j), j = 1, 2$.

2.2 Integral equations for anisotropic bounded plate with holes and cracks

Let us write a general solution of the problem based on [12, 15] through the Lekhnitskii potentials in the form

$$\begin{aligned}\Phi(z_1) &= \int_L [u'\Phi_1(z_1,t_1) + v'\Phi_2(z_1,t_1)]ds \\ &+ \int_\Gamma \left[g_1'\Phi_1(z_1,t_1) + g_2'\Phi_2(z_1,t_1)\right]ds + \Phi_S(z_1) + \Phi_\Delta(z_1), \\ \Psi(z_2) &= \int_L [u'\Psi_1(z_2,t_2) + v'\Psi_2(z_2,t_2)]ds \\ &+ \int_\Gamma \left[g_1'\Psi_1(z_2,t_2) + g_2'\Psi_2(z_2,t_2)\right]ds + \Psi_S(z_2) + \Psi_\Delta(z_2),\end{aligned} \tag{6}$$

where $L = L_0 + L_1 + ... + L_J$, $\Gamma = \Gamma_1 + \Gamma_2 + ... + \Gamma_K$, s is an arc coordinate, and $\Phi_\Delta(z_1)$ and $\Psi_\Delta(z_2)$ are the known functions, which are determined by the following formulas:

$$\Phi_\Delta(z_1) = \int_L [X_L\Phi_3(z_1,t_1) + Y_L\Phi_4(z_1,t_1)]ds,\ \Psi_\Delta(z_1) = \int_L [X_L\Psi_3(z_1,t_1) + Y_L\Psi_4(z_1,t_1)]ds, \tag{7}$$

$$\Phi_j = \frac{A_j}{t_1 - z_1},\ \Psi_j = \frac{B_j}{t_2 - z_2}, \tag{8}$$

$$A_1 = -\frac{is_1}{2\pi\Delta_1},\quad A_2 = \frac{i}{2\pi\Delta_1}, A_3 = -\frac{ip_1}{2\pi\Delta_1},\quad A_4 = -\frac{iq_1}{2\pi\Delta_1},$$

$$B_1 = -\frac{is_2}{2\pi\Delta_2},\quad B_2 = \frac{i}{2\pi\Delta_2}, B_3 = -\frac{ip_2}{2\pi\Delta_2},\quad B_4 = -\frac{iq_2}{2\pi\Delta_2}.$$

Here, u', v' are the values of the derivatives of the displacements with respect to the arc coordinate at the boundary of the plate and holes,
$g_1 = u^+ - u^-, g_2 = v^+ - v^-$ are the displacements discontinuity at the cracks, $u^\pm, v^\pm$ are limit values of displacements in the approach to the section at the left and the right relative to the selected direction, and the potentials Φ_S, Ψ_S correspond to the concentrated forces and have the form [12]:

$$\Phi_S(z_1) = \frac{i}{2\pi\Delta_1}\sum_{j=1}^{M}(p_1X_j + q_1Y_j)\frac{1}{z_1 - z_{1j}},\ \Psi_S(z_2) = \frac{i}{2\pi\Delta_2}\sum_{j=1}^{M}(p_2X_j + q_2Y_j)\frac{1}{z_2 - z_{2j}}, \tag{9}$$

where in $z_{kj} = a_j + s_k b_j, j$ = 1, 2, and k = 1, 2.

Note that when the boundary is traction free, then $\Phi_\Delta = \Psi_\Delta = 0$.

Let us substitute the potentials (6) into the formulas (4) for projections of stress vectors determined at the boundaries path L and Γ. Using Plemelj-Sokhotski formula, we obtain a system of integral equations [12, 15]:

$$\begin{aligned}&\int_L [u'(s)Q_1(Z,T) + v'(s)Q_2(Z,T)]ds \\ &+ \int_\Gamma [g_1'(s)Q_1(Z,T) + g_2'(s)Q_2(Z,T)]ds = Q(Z), Z \in L \cup \Gamma,\end{aligned} \tag{10}$$

where $Q_j(Z,T)$ are stress vectors q_L at point Z with coordinates (x,y). $L\cup\Gamma$, the stress vector is determined by the formula (4) accordingly through complex potentials $\Phi_j(z_1,t_1), \Psi_j(z_2,t_2),\ j = 1, 2$; T is a point with coordinates (ξ,η), which belongs to the contour $L\cup\Gamma$; $Q(Z) = Q_L(Z) - Q_S(Z) - Q_\Delta(Z)$ with $Z\in L$ and $Q(Z) = Q_T(Z) - Q_S(Z) - Q_\Delta(Z)$ with $Z\in\Gamma$; $Q_L = X_L + iY_L$; and $Q_m = X_m + iY_m$, where $X_m = 2\mathrm{Re}\left[s_1\Phi_m(z_1)z_1' + s_2\Psi_m(z_2)z_2'\right]$, $Y_m = -2\mathrm{Re}\left[\Phi_m(z_1)z_1' + \Psi_m(z_2)z_2'\right]$ and

$$m = S,\ \Delta.$$

Using the results [12], we obtained that the unknown functions u', v' at the boundary of each of the holes $L_j, j = 0, 1, \ldots, J$ in representation (6) are defined up to a summand $\tilde{u}' = -\omega_j dy/ds, \tilde{v}' = \omega_j dx/ds$, where ω_j are arbitrary constants. At numerical solution of the problem, the constants $\omega_j, j = 0, \ldots, J$ are to be necessarily fixed. In addition, to ensure the displacement continuity condition, it is necessary to impose the following conditions on unknown functions:

$$\int_{L_j} u' ds = 0, \int_{L_j} v' ds = 0, j = 0, ..., J; .$$
$$\int_{\Gamma_j} g_1' ds = 0, \int_{\Gamma_j} g_2' ds = 0, j = 1, ..., K \tag{11}$$

Let us consider a problem-solving equation (10) for the case of one hole and a crack. Let us assume that the contour on which the crack is placed is described parametrically in the form $x = \alpha_\Gamma(\tau), y = \beta_\Gamma(\tau), \ -1 \le \tau \le 1$, and the equation of the boundary hole is described in the form $x = \alpha_L(\theta), y = \beta_L(\theta), 0 \le \theta < 2\pi$.

Let us assume the representation for the displacement discontinuity at the cracks:

$$g_1' s' = \frac{dg_1}{d\tau} = \frac{U_\Gamma(\tau)}{\sqrt{1-\tau^2}}, \ g_2' s' = \frac{dg_2}{d\tau} = \frac{V_\Gamma(\tau)}{\sqrt{1-\tau^2}}.$$

Let us replace the integrals with Lobatto-type quadrature formulas [15], and the integrals at the boundaries of the holes replaced by the quadrature of a rectangle, which, for periodic functions, are Gauss quadrature-type formulas [12]. Then we obtain the system of equations:

$$H \sum_{k=1}^{N_O} \left(q_{\nu k}^{(1)} U_k^L + q_{\nu k}^{(2)} V_k^L \right) + \sum_{m=1}^{N_\Gamma} C_m \left(q_{\nu m}^{(1)} U_m^\Gamma + q_{\nu m}^{(2)} V_m^\Gamma \right) = q_\nu, \nu = 1, ..., N_O + N_\Gamma - 1, \tag{12}$$

where $q_{\nu k}^{(j)} = Q_j(Z_\nu, T_k^L), p_{\nu m}^{(j)} = Q_j(Z_\nu, T_m^\Gamma), q_\nu = Q_L(Z_\nu) - Q_S(Z_\nu) - Q_\Delta(Z_\nu)$,
$U_k^L = u'(x_k^L, y_k^L) s_k', V_k^L = v'(x_k^L, y_k^L) s_k', U_m^\Gamma = g_1'(x_m^\Gamma, y_m^\Gamma) s_m', V_m^\Gamma = g_2'(x_m^\Gamma, y_m^\Gamma) s_m'$,
$x_k^L = \alpha_L(\theta_k), \ y_k^L = \beta_L(\theta_k), \theta_k = Hk, \tilde{\theta}_n = \theta_n - H/2, H = 2\pi/N_O$,
$x_m^T = \alpha_\Gamma(\tau_m), \ y_m^T = \beta_\Gamma(\tau_m), \tau_m = -\cos(\pi_N(m-1)), m = 1, ..., N_\Gamma; C_m = \pi_N$ at $m \ne 1$ and $m \ne N_\Gamma; C_1 = C_{N_\Gamma} = 0, 5\pi_N; \pi_N = \frac{\pi}{N_\Gamma - 1}; T_k^L$ is a point with coordinates (x_k^L, y_k^L), T_m^Γ is a point with coordinates (x_m^Γ, y_m^Γ), Z_ν is a point with coordinates $(\tilde{x}_\nu^L, \tilde{y}_\nu^L)$ with $1 \le \nu \le N_L$, and is a point with coordinates $\left(\tilde{x}_{\nu - N_L}^\Gamma, \tilde{y}_{\nu - N_L}^\Gamma\right)$ with $N_L < \nu \le N_L + N_\Gamma - 1$, where $\tilde{x}_k^L = \alpha_L(\tilde{\theta}_k), \ \tilde{y}_k^L = \beta_L(\tilde{\theta}_k), \ \tilde{\theta}_n = \theta_n - H/2, \tilde{x}_m^\Gamma = \alpha_\Gamma(\tilde{\tau}_m), \ \tilde{y}_m^\Gamma = \beta_\Gamma(\tilde{\tau}_m), \tilde{\tau}_m = -\cos[\pi_N(m - 0,5)]$.

We obtain the additional equation of system (12) from condition (11)

$$\sum_{k=1}^{N_\Gamma} C_k \left(U_k^\Gamma + i V_k^\Gamma \right) = 0. \tag{13}$$

Analogously to [12], we should remove three equations with $1 \le \nu \le N_L$ and add the following three equations to the received incomplete system:

$$\sum_{k=1}^{N} U_k^L = 0, \sum_{k=1}^{N} V_k^L = 0, U_m^L = 0, 1 \le m \le N \tag{14}$$

The first two equations follow from the displacements continuity conditions (9). The last equation is obtained when fixing an arbitrary constant ω.

The system of Eqs. (12)–(14) is generalized in the case of hole and crack system in the same way as it was done in [12].

3. Stresses in circular samples with cracks under the action of concentrated forces

Let us consider the circular composite plate with radius a, which is weakened by a central crack with the half-length L. The plate is stretched by the concentrated forces $\pm P$ applied at points $(0, \pm y_0)$. Destruction of the plate is happening when the stress intensity factors (SIFs) reach a certain limit value. Therefore, when we calculated the limiting loads, we considered the SIF which explicitly takes into account the length of the cracks. Due to this, we performed calculations of the relative SIF $K_a = \frac{K_I\sqrt{a}}{P\sqrt{\pi}}$ with different relative distances $\alpha = y_0/a$, depending on the half-length of the crack, which is divided into a.

Calculations are made for the composite plates with elastic constants shown in **Table 1**.

The results of the calculations for the plate made of an EF material (with a small degree of anisotropy) with the maximum stiffness in the direction of the OX axis are shown in **Figure 3**.

In general, the character of the distribution for an EF material is not significantly different from that of an isotropic material. It is necessary to increase monotonically the load for a stable growth of the crack when the distances of forces to a crack are smaller than 0, 2a. With $y_0/a = 0, 3$, the stable growth of cracks (without jumping) will occur at $L/a > 0, 2$; with $y_0/a = 0, 4$ if $L/a > 0, 25$; and with $y_0/a = 0, 5$ if $L/a > 0, 35$. At greater distances to forces, after reaching the corresponding level of values of traction, the circle fractures. For a case where the crack is perpendicular to the direction with the maximum stiffness of the material, the SIF is slightly increasing, especially at greater distances to forces.

Material	E_1 (GPa)	E_2 (GPa)	G_{12} (GPa)	ν_{21}	ν_{12}
LU	10.8	96.0	2.61	0.21	0.024
EF	21	32.8	5.7	0.21	0.134

Table 1.
Elastic constants of LU and EF materials.

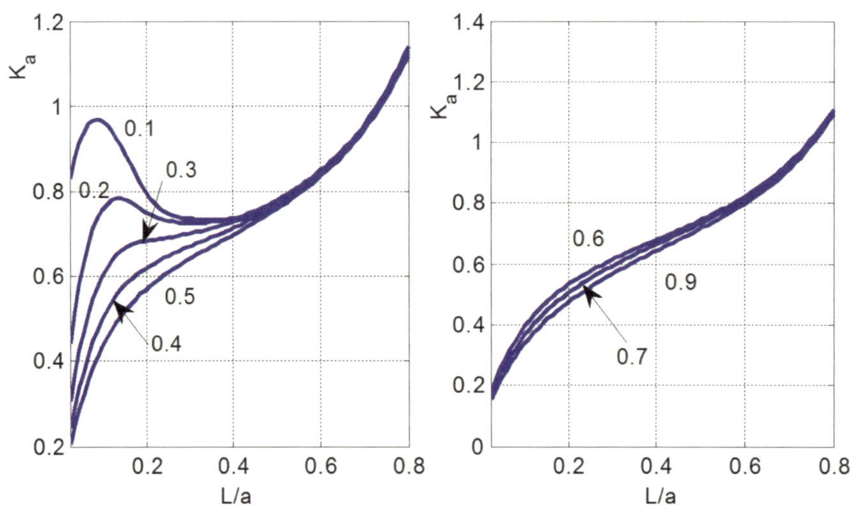

Figure 3.
Relative SIF for a circular plate made of an EF material: a direction with maximum stiffness is parallel to the OX axis.

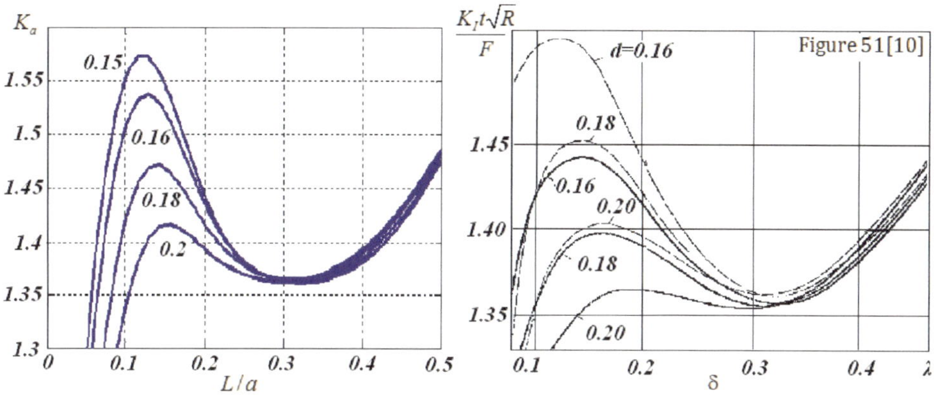

Figure 4.
Relative SIF for isotropic material.

For a weakly anisotropic material, the incline of the crack to the main axis of orthotropy had little effect on the SIF K_I, and the SIF K_{II} is practically absent.

The calculations have shown that for the case of placing the crack in parallel to the direction with the maximum stiffness of the material, the above set of specifics of the SIF remain unchanged for substantially anisotropic LU-1 material. When the crack is placed perpendicular to the direction with maximum stiffness, stable crack growth occurs of small cracks ($L/a \leq 0, 1$) when the relative distance to forces is $y_0/a < 0, 2$. Moreover, for these cases, the fracture is spasmodic. In all other cases, the SIF increases monotonically with increasing crack length. Hence, the plate fractures completely after reaching forces of critical value.

Testing the developed algorithm is conducted for the case of isotropic plate with $\nu = 0, 4488$ and $y_0/a = 0, 15; 0, 16; 0, 18; 0, 2$. The calculation results of the relative SIF $K_a = \frac{K_I\sqrt{a}}{P}$ are shown in **Figure 4**.

On the right are shown figures from the book [10]. Such calculations were also performed for the same material, and here the corresponding relative SIF is represented by dashed lines. It is seen that the results obtained by different methods coincide.

4. Determination of working intervals of crack lengths at circular samples

Two types of samples are used in experimental studies of crack fracture resistance [10]. The first is a sample for which the SIF grows monotonically with the growth of the crack. In the second type, the range of cracks' lengths is selected in such a way that the SIF K_I is practically constant. Hence, it is varied in this range from the mean value to the small value (~2–4%). This range of crack lengths is called working. The samples of the second type are particularly suitable for conducting experimental studies including a wide range of problems in the area of destruction. In particular, with the constant force factor (the SIF is constant) in such samples, the possibility of an effective study of the rate of growth of fatigue cracks with cyclic loads, the study of crack fracture resistance depending on the influence of working environment, etc. arises.

Based on the studies in the literature for isotropic material, it has been established [10] that the range of working lengths is most favorable with $\alpha = 0, 18$, although allowed, and $\alpha = 0, 16; 0, 2$, where $\alpha = y_0/a$. With $0 < \alpha < 0, 2237$ the SIF K_I increases from zero to a certain maximum (depending on α), then falls to a minimum, and then increases monotonously. With α>0, 2237 the SIF increases monotonically with the increasing length of the crack. We note that such conclusions are valid for material with a Poisson ratio $\nu = 0, 4488$.

Since at big lengths of cracks L/a>0, 5 the SIF depends little on α, then for the first type of samples $\alpha = 0, 65$ is taken.

Based on the obtained results, let us perform a similar study of samples of two types of composite materials.

For samples of the second type, we perform calculations only for small ratios (at $\alpha \sim 0, 2$). The results of calculations of the SIF in a circular isotropic sample with $\nu = 1/3$ are shown in **Figure 5**. Here and further is assumed that the relative value of the SIF is equal to $K_a = \frac{K_I\sqrt{a}}{P\sqrt{\pi}}$ and the parameter value α is indicated near the curves.

Figure 5 shows that the range with few changed SIF is necessary to determine in the vicinity of the lengths of the cracks with $L/a \sim 0, 25$.

Similar results for the plate made of an EF material are shown in **Figures 6** and 7. Here two cases are considered: the crack is parallel or perpendicular to the direction in which the stiffness of the material is maximal.

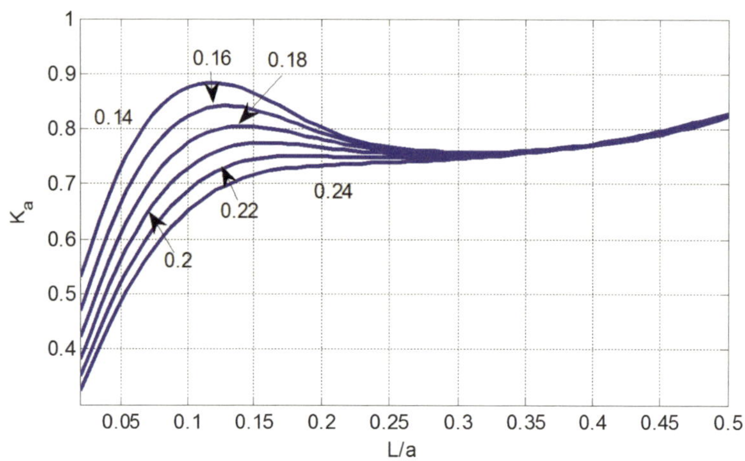

Figure 5.
Relative SIF for isotropic material.

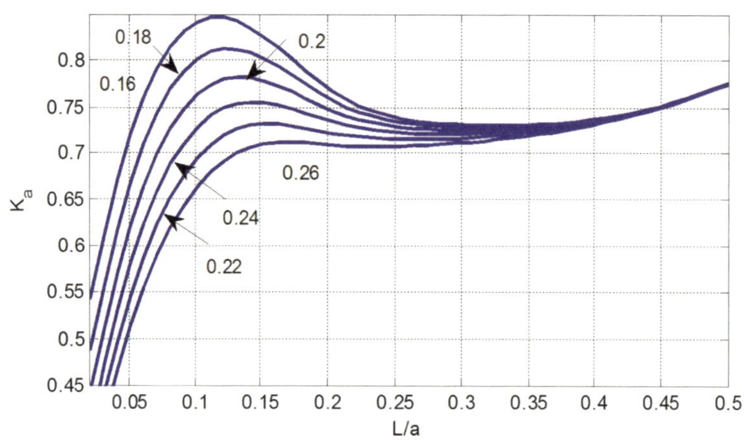

Figure 6.
Relative SIF for an EF material: the crack is parallel to the direction of greater stiffness of the material.

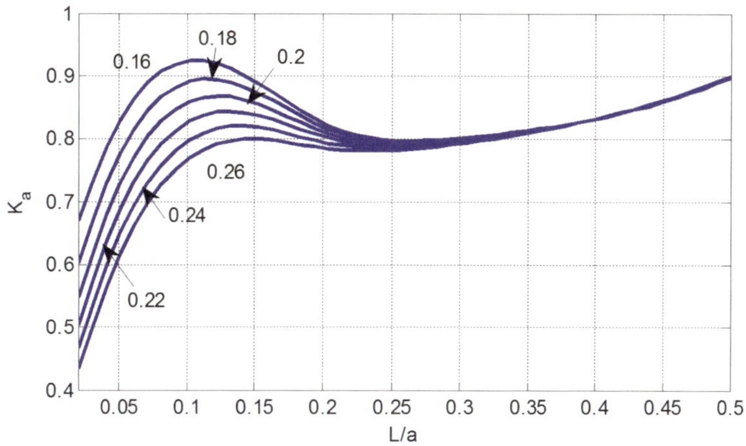

Figure 7.
Relative SIF for an EF material: the crack is perpendicular to the direction of greater stiffness of the material.

From the given data, it is seen that the SIF for an EF material is bigger in the case when the crack is perpendicular to the direction of maximum stiffness of the material.

The following conclusions are made based on the data of the calculations: for the crack that is parallel to the direction of maximum stiffness of the material, the minimum deviations from the constant SIF (with an error of not more than 2%) are achieved on the ranges of lengths of cracks with relative dimensions with $\Delta < 0, 21$ with $\alpha = 0, 24 - 0, 26$, where $\Delta = (L_2 - L_1)/a$. The biggest range ($\Delta = 0, 27$) with SIF values close to constants with an error 2, 1% is achieved with $\alpha = 0, 24$. For the crack that is perpendicular to the direction of maximum stiffness of the material, the range with SIF values close to constants is reduced. Moreover, the distance of force application needs to be increased. The biggest range ($\Delta = 0, 27$) with SIF values close to constants with an error 3, 2% is achieved with $\alpha = 0, 24$.

Let us consider the case of samples of the first type 1. For them, the forces are selected that are distant from the crack. For isotropic materials, as a rule, $\alpha = 0, 65$ is taken. The above-obtained results of the calculations show that the same distance can be chosen for the composite materials with a crack parallel to the principal axes of the orthotropy.

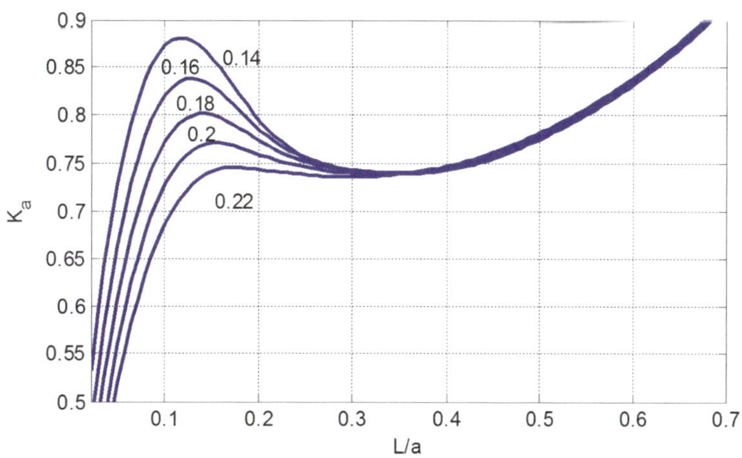

Figure 8.
Relative SIF for a square isotropic sample.

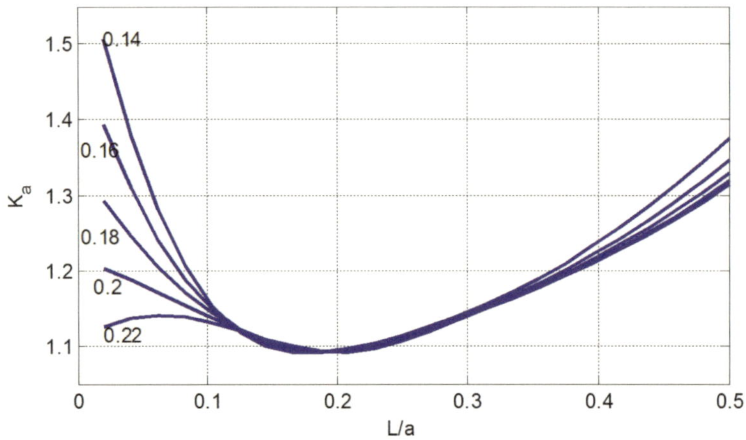

Figure 9.
LU material, square sample, horizontal crack.

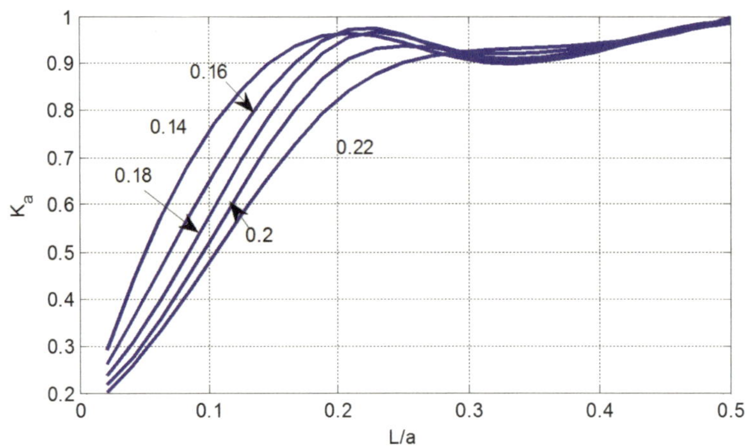

Figure 10.
LU material, square sample, diagonal crack.

To compare the effect of a sample shape, similar calculations are made for a square plate with a crack. The results of calculations for such an isotropic sample that are similar to the results of calculations for a circle are shown in **Figure 8**.

The conclusion is made based on the comparison of **Figures 5** and **8** that with small distances of forces from cracks, the shape of the sample has little effect on the SIF.

Similar results of calculations for LU material are shown in **Figure 9** for a horizontal crack and in **Figure 10** for a diagonal crack.

5. Determination of the SSS of samples under the action of the tractions applied to the hole's boundary

Let us apply the developed algorithm to study a square plate with a half-side a; weakened by a central crack with a half-length L (**Figure 11**), the edges are not loaded.

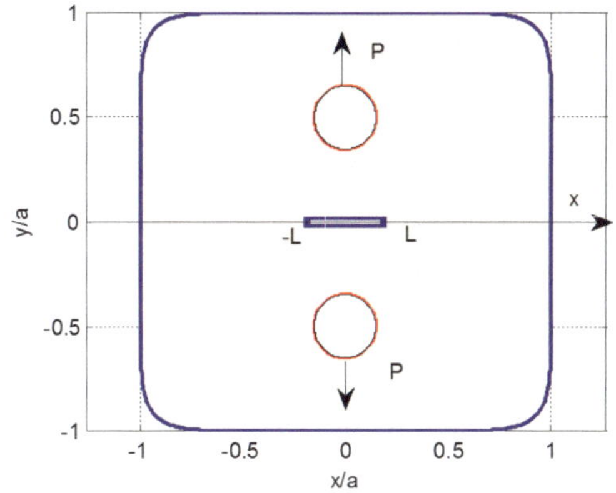

Figure 11.
Scheme of the sample.

Two identical circular holes of radius R, the centers of which are located at points $(0, \pm c)$, are created for stretching in a plate. It was assumed that the load was applied to the boundary of the circular holes. Using [10], we accept that the forces act normally on the domain $|\theta - \theta_c| < \gamma$ and are given on it in the form

$$p = P\frac{(\theta - \theta_c)^2}{4R(\sin\gamma - \gamma\cos\gamma)}, \tag{15}$$

where θ is angle coordinate on each of the holes, θ_c is angle coordinate of the middle of the domain, and P is the principal vector applied to the domain of forces, which is directed from the center of the hole at an angle θ_c.

At first, for the purpose of testing the algorithm, the calculations are performed for the case of a localized load at $\gamma = \pi/32$ and $\alpha = c/a = 0, 2; 0, 4; 0, 6$, with $\theta_c = \pi/2$ for the upper hole and $\theta_c = -\pi/2$ for the lower hole (thus the stretching of the plate in the direction of the OY axis is considered). The relative SIF $Y = \frac{K_I}{P}\sqrt{\pi L}$ for an isotropic material with a Poisson ratio $\nu = 1/3$ at different crack lengths at $R/a = 0, 1$ is calculated and given in **Table 2**.

The values of relative SIF Y_S obtained by another method in [10] for the case of stretching by concentrated forces (i.e., with $\gamma \to 0$) are given in the same table.

L/a	Y	Y_S	Y	Y_S	Y	Y_S
	$\alpha = 0.2$		$\alpha = 0.4$		$\alpha = 0.6$	
0.1	0.175	0.177	0.248	0.247	0.271	0.270
0.2	0.721	1.720	0.583	0.581	0.568	0.565
0.3	1.164	1.159	0.958	0.953	0.894	0.888
0.4	1.503	1.495	1.323	1.314	1.237	1.227
0.5	1.819	1.805	1.675	1.663	1.588	1.575
0.6	2.156	2.131	2.034	2.015	1.954	1.935
0.7	2.552	2.505	2.438	2.406	2.367	2.334
0.8	3.112	2.995	2.999	2.917	2.937	2.854
0.9	4.279	—	4.178	—	4.131	—

Table 2.
Relative SIFs for a square sample, isotropy.

As you can see, the results for these cases were close, with the exception of the values $L/a = 0, 8$. Some differences in them are due to different localizations of applied tractions.

The results of the calculations of SIF are given in **Table 3** for the case $R \to 0,\ \gamma = 0$, that is, the case is considered when forces are applied at the center of the holes and more smoothly applied efforts (at $\gamma = \pi/8$).

The results for a circular sample with $\gamma = \pi/8$ are shown (**Figure 12**) in the same table.

The following conclusions are made based on **Tables 2** and **3**: SIF does not differ significantly in the case of distributed loads with different degrees of localization, SIF increases somewhat with the growth of the domain of action of tractions, and SIFs are bigger at small crack lengths at point action of tractions (at $R \to 0$) and at all lengths with $\alpha > 0, 6$. SIFs in a circular sample are bigger than in a square one under the same load conditions.

Similar results for a square sample made from a LU material are given in **Table 4**. Here, the relative SIFs K_a in which the crack length is explicitly taken into

	$\gamma = \pi/8$			$R \to 0,\ \gamma = 0$			$\gamma = \pi/8$ (circle)		
L/a	$\alpha = 0.2$	0.4	0.6	$\alpha = 0.2$	0.4	0.6	$\alpha = 0.2$	0.4	0.6
0.1	0.186	0.254	0.275	0.727	0.462	0.366	0.2069	0.2637	0.2767
0.2	0.732	0.593	0.576	1.073	0.851	0.717	0.7563	0.6096	0.5786
0.3	1.172	0.969	0.905	1.278	1.167	1.049	1.1914	0.9910	0.9101
0.4	1.510	1.335	1.250	1.489	1.452	1.370	1.5415	1.3678	1.2655
0.5	1.827	1.689	1.605	1.742	1.744	1.694	1.8964	1.7516	1.6483
0.6	2.167	2.050	1.974	2.048	2.067	2.039	2.3113	2.1770	2.0811
0.7	2.569	2.459	2.391	2.436	2.459	2.440	2.8332	2.6963	2.6106
0.8	3.139	3.029	2.971	3.008	3.027	3.011	3.5616	3.4216	3.3492
0.9	4.322	4.225	4.181	4.214	4.225	4.209	4.8896	4.7662	4.7132

Table 3.
Relative SIFs for a square and circular sample, isotropy.

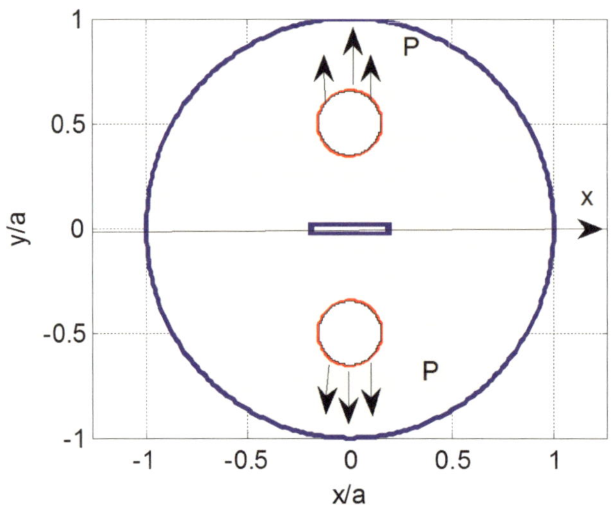

Figure 12.
Circular sample with holes and a crack.

		LU (Ox)			LU (Oy)	
L/a	$\alpha = 0.2$	0.4	0.6	$\alpha = 0.2$	0.4	0.6
0.1	0.2338	0.2633	0.2764	0.4182	0.478	0.538
0.2	0.5187	0.4276	0.4059	1.0504	1.0073	0.9911
0.3	0.5952	0.5259	0.4978	1.1416	1.1306	1.1327
0.4	0.6202	0.5791	0.5595	1.2268	1.2138	1.2163
0.5	0.6409	0.6158	0.6046	1.33	1.3094	1.31
0.6	0.6711	0.6539	0.6479	1.4615	1.4313	1.4304
0.7	0.7252	0.7117	0.7085	1.6257	1.5829	1.5809
0.8	0.8392	0.8284	0.8267	1.7876	1.7411	1.7386
0.9	1.1482	1.1419	1.1409	1.9405	1.8846	1.8831

Table 4.
Relative SIFs K_a for square sample, LU material.

		LU (Ox)			LU (Oy)	
L/a	$\alpha = 0.2$	0.4	0.6	$\alpha = 0.2$	0.4	0.6
0.1	0.2363	0.2679	0.2806	0.4192	0.4799	0.5407
0.2	0.5269	0.4369	0.4138	1.0594	1.0163	0.9995
0.3	0.6108	0.5418	0.5112	1.1602	1.1477	1.1466
0.4	0.6467	0.605	0.5814	1.2588	1.2422	1.2389
0.5	0.683	0.6558	0.6396	1.3795	1.3531	1.3459
0.6	0.7341	0.7131	0.702	1.5354	1.4974	1.4888
0.7	0.8131	0.7948	0.7879	1.7482	1.6965	1.6905
0.8	0.9524	0.9376	0.9342	2.0507	1.9973	1.9977
0.9	1.2791	1.2726	1.2704	2.592	2.528	2.5346

Table 5.
Relative SIFs for a circular sample, LU material.

account and the case where the crack is parallel to the direction of bigger stiffness (data on the left) and is perpendicular to it (data on the right) are given.

Similar results for a circular sample for the same material are given in **Table 5**.

6. Determination of the SSS of samples loaded with concentrated forces at the boundary (compression test)

The methods of studied crack fracture resistance based on sample compression in experimental practice are widely used.

The direct application of the abovementioned variant of the method of integral equations for the case when the concentrated forces act on the boundary of the domain is associated with significant errors, because unknown functions in the vicinity of the points of application of forces have a singularity. Due to this, it is necessary to separate a singular part in the solution for a more precise solution of this type of task.

6.1 Determining a singular part of the solution of this problem

Let us consider a point of the plate boundary z_0, in which the concentrated force (X, Y) is applied. A singular part of the solution (of Lekhnitskii potentials) will be the same as in the half-plane, whose boundary is tangent to the plate at the point of action of the concentrated force. Let us mark this angle through φ and potentials for the half-plane through $\Phi_0(z_1)$, $\Psi_0(z_2)$.

Let us consider at first a half-plane $y < 0$, which is loaded with force (X, Y) at an arbitrary point x_0 on the boundary. The Lekhnitskii potentials for this half-plane will be [17]

$$\Phi_0 = \frac{A'}{z_1 - x_0}, \Psi_0 = \frac{B'}{z_2 - x_0},$$

where

$$A' = -\frac{X + s_2 Y}{2\pi i(s_1 - s_2)}, B' = \frac{X + s_1 Y}{2\pi i(s_1 - s_2)} .$$

It can be shown [17] that the half-plane also corresponds to the Φ_0, Ψ_0 potentials, whose boundary passes through the point $(x_0, 0)$ and is inclined at an arbitrary angle under the action of the same force. Let us consider a semicircle with the center at point $(x_0, 0)$ of radius ρ_0, which belongs to the half-plane. It is easy to show that the principal vector of all forces applied to the arc of the semicircle is equal to $(-X, -Y)$. This proves that the case of loading by the concentrated force of the half-plane corresponds to the potentials Φ_0, Ψ_0.

Let us now consider a bounded plate occupying the domain D. The self-balanced concentrated forces (X_j, Y_j), $(j = 1, ..., J)$. are applied to the boundary of this domain at points $z_j = x_j + iy_j$. Let us represent the complex potentials in the form

$$\Phi(z_1) = \Phi_0(z_1) + \Phi_\Delta(z_1), \Psi(z_1) = \Psi_0(z_1) + \Psi_\Delta(z_1), \tag{16}$$
$$\Phi_0(z_1) = \sum_{j=1}^{J} \frac{A'_j}{z_1 - z_{1j}}, \Psi_0(z_2) = \sum_{j=1}^{J} \frac{B'_j}{z_2 - z_{2j}},$$

where $z_{1j} = x_j + s_1 y_j, z_{2j} = x_j + s_2 y_j$. Here the coefficients A'_j, B'_j are determined based on expressions for A', B' by the substitution of X and Y on X_j and Y_j, respectively. By substituting formulas (16) into boundary conditions, we obtain the boundary problem for obtaining the introduced complex potentials at $(x, y) \in L$:

$$\begin{aligned}&(1 + is_1)z_1'\Phi_\Delta(z_1) + \overline{(1 + is_1)z_1'\Phi_\Delta(z_1)} + \\ &+(1 + is_2)z_2'\Psi_\Delta(z_2) + \overline{(1 + is_2)z_2'\Psi_\Delta(z_2)} = -i(X + iY)_0,\end{aligned} \tag{17}$$

where L is the boundary of domain D:

$$i(X + iY)_0 = (1 + is_1)z_1'\Phi_0(z_1) + \overline{(1 + is_1)z_1'\Phi_0(z_1)} + (1 + is_2)z_2'\Psi_0(z_2) + \overline{(1 + is_2)z_2'\Psi_0(z_2)}.$$

It is easy to show that the right-hand side of formula (17) is a continuous and limited function, and therefore the introduced complex potentials with an index Δ are continuous and limited in the vicinity of the points of application of forces. In this regard, the above-developed numerical algorithm based on BIEM can be used to determine these potentials.

6.2 Calculation of the SIF for a rectangular sample with compression

Let us consider a square plate with half-side a, which contains a diagonal central vertical crack with half-length L. The crack fracture resistance of such a sample is determined based on compression by force R applied in vertical direction. The tips of the angles, in the vicinity of which forces are applied, can be cut off. Due to this, a sample is considered whose tops have coordinates

$$z_1 = c, z_2 = ic + z_p, z_3 = ic + z_m, z_4, z_5 = \overline{z_3}\, z_6 = \overline{z_2},$$

where in $c = \sqrt{2}a, z_p = h - ih, z_m = -h - ih,$ h is the height of the cut triangle and a is the half-side of the square.

The calculations are performed at $h = c/8$; moreover, all tops are rounded by the arcs of the circle of the radius $a/10$ (the shape of the sample—**Figure 13**).

Calculation of the relative SIF $K_a = \frac{K_I\sqrt{a}}{P\sqrt{\pi}}$ for isotopic material and composites of EF and LU with different directions of the orthotropic axis are given in **Table 6**. The angle between the crack and the direction with the maximum stiffness of the material is indicated in brackets.

The table shows a significant effect on the SIF of the placement of the crack relative to the axis with the maximum stiffness of the material. In particular, for

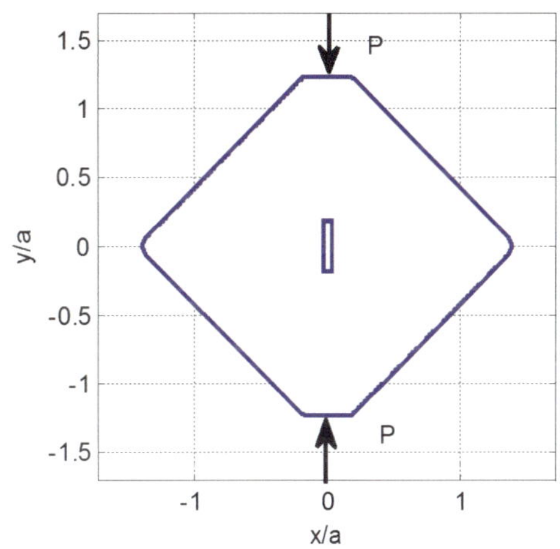

Figure 13.
Sample view.

L/a	Isotr.	EF (90°)	EF (0°)	LU (90°)	LU (0°)
0.1	0.0824	0.0579	0.0727	0.0226	0.0743
0.2	0.1203	0.085	0.1073	0.0332	0.1182
0.3	0.1557	0.1106	0.1407	0.0434	0.1663
0.4	0.1927	0.1384	0.1776	0.0547	0.2217
0.5	0.2354	0.1709	0.2208	0.068	0.2865
0.6	0.2889	0.2107	0.2737	0.0846	0.3643
0.7	0.3544	0.2609	0.3406	0.1059	0.4609
0.8	0.4418	0.3273	0.4301	0.1343	0.5872

Table 6.
Relative SIF K_a when compressing a sample with a diagonal crack.

L/a	Isotr.	EF (0°)	EF (90°)	LU (0°)	LU (90°)
0.1	0.2585	0.2886	0.3499	0.2956	0.7572
0.2	0.3696	0.4065	0.4874	0.4114	0.9247
0.3	0.4617	0.4963	0.5884	0.4936	1.0104
0.4	0.5494	0.5752	0.6766	0.5594	1.1008
0.5	0.641	0.6534	0.7668	0.6185	1.2159
0.6	0.7434	0.7397	0.8705	0.6789	1.364
0.7	0.8641	0.842	0.9973	0.7472	1.5542
0.8	1.0129	0.9704	1.1585	0.8319	1.8021

Table 7.
Relative SIFs when stretching a sample with a diagonal crack.

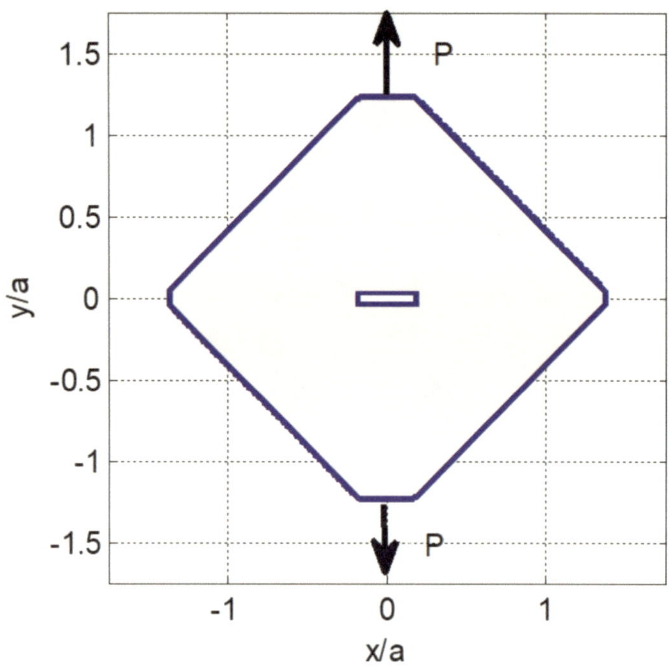

Figure 14.
Sample view.

cracks parallel to the maximum stiffness direction, the SIFs appeared to be significantly larger than those returned for 90°. The difference between the SIFs for these two directions is increasing for a substantially anisotropic LU material.

Table 7 shows the results of calculations for the case of stretching the same sample with a horizontal crack (**Figure 14**).

Based on the comparison of data from **Tables 6** and 7, it follows that, unlike the case of compression, the SIF with stretching is larger for cracks that are perpendicular to the direction with maximum stiffness of the material.

7. Conclusions

An algorithm for calculating stresses at cracks in bounded plate with holes of various shapes due to concentrated forces or distributed forces at its boundary has developed. The solution of integral equations is performed by quadrature Gauss-type formulas for regular and singular integrals.

The research of stresses at cracks in the samples which are used in experimental studies of crack fracture resistance was performed.

The calculation of the stresses at cracks in samples of various forms is performed, in which ones' experimental research are performed. To study the crack fracture resistance of composite samples, the optimal distances from the central crack to the forces at which the SIF increases monotonously with increasing crack length are determined. In particular, for square samples with a half-side a, forces should be placed at a distance of $0.6a$–$0.7a$ from the crack. For the experimental study of the growth rate of fatigue cracks, there are definite ranges of lengths of cracks for which SIFs are practically constant values. At the same time, the distances are determined at which it is expedient to apply forces. The problem for studying samples with cracks with compression is considered.

List of symbols with explanations

s is an arc coordinate.

ν is a Poisson ratio.

$\varepsilon_x, \varepsilon_y, \gamma_{xy}$ are strains.

$\sigma_x, \sigma_y, \tau_{xy}$ are stresses.

α_{ij} are elastic compliances which are included in the Hooke's law.

$\Phi(z_1), \Psi(z_2)$ are Lekhnitskii complex potentials.

Φ_S, Ψ_S are the potentials which correspond to the concentrated forces.

u', v' are the values of the derivatives of the displacements with respect to the arc coordinate at the boundary of the plate and holes.

$u^{\pm}, v^{\pm}$ are limit values of displacements in the approach to the section at the left and the right relative to the selected direction.

$\vec{q_\Gamma}$ is the stress vector $q_\Gamma(z) = X_\Gamma + iY_\Gamma$ at path Γ.

K_I, K_{II} are the stress intensity factors (SIFs).

K_a, Y, Y_S are the relative SIFs.

P is the principal vector.

Author details

Olesia Maksymovych[1,2]* and Adam Podhorecki[2]

1 Department of Welding Manufacture, Diagnostics and Restoration of "Lviv Polytechnic National University", Lviv, Ukraine

2 Faculty of Civil, Architecture and Environmental Engineering, University of Technology and Life Science, Bydgoszcz, Poland

*Address all correspondence to: olesyamax@meta.ua

References

[1] Tsukrov I, Kachanov M. Effective moduli of an anisotropic material with elliptical holes of arbitrary orientational distribution intern. International Journal of Solids and Structures. 2000;**37**(41): 5919-5941. DOI: 10.1016/S0020-7683(99)00244-9

[2] Scholar NA. Anisotropic Analysis Using Boundary Elements. Computational Mechanics Publications; 1994, 142 p

[3] Brebbia CA, Dominguez J. Boundary Elements: An Introductory Course. McGraw Hill Book Co., Computational Mechanics Publications. 1989

[4] Sollero P, Aliabadi MH. Fracture mechanics analysis of anisotropic plates by the boundary element method. International Journal of Fracture. 1993; **64**:269-284. DOI: 10.1007/BF00017845

[5] Tan CL, Gao YL. Boundary element analysis of plane anisotropic bodies with stress concentrations and cracks. Composite Structures. 1992;**20**:17-28. DOI: 10.1016/0263-8223(92)90008-Z

[6] Aliabadi MH, Brebbia CA. Boundary element formulations in fracture mechanics: A review. WIT Transactions on Engineering Sciences. 1970;**13**:17

[7] Garcıa F, Sáez A, Domınguez J. Traction boundary elements for cracks in anisotropic solids. Engineering Analysis with Boundary Elements. 2004;**28**(6):667-676. DOI: 10.1016/j.enganabound.2003.08.005

[8] Tu CH, Chen CS, Yu TT. Fracture mechanics analysis of multiple cracks in anisotropic media. International Journal for Numerical and Analytical Methods in Geomechanics. 2011;**35**(11): 1226-1242. DOI: 10.1002/nag.953

[9] Yang S, Yuan F-G. Kinked crack in anisotropic bodies. International Journal of Solids and Structures. 2000;**37**: 6635-6682. DOI: 10.1016/S0020-7683(99)00222-X

[10] Savruk MP, Osiv PN, Prokopchuk IV. Numerical Analysis in Plane Problems of Cracks Theory. Kiev: Naukova Dumka; 1989. p. 248

[11] Savruk MP, Kazberuk A. Stress Concentration at Notches. Cham: Springer; 2017

[12] Maksymovych O, Illiushyn O. Stress calculation and optimization in composite plates with holes based on the modified integral equation method. Engineering Analysis with Boundary Elements. 2017;**83**:180-187. DOI: 10.1016/j.enganabound.2017.06.009

[13] Maksymovych O, Jaroszewicz J. Determination of stress state of anisotropic plates with rigid inclusions based on singular integral equations. Engineering Analysis with Boundary Element. 2018;**95**:215-221. DOI: 10.1016/j.enganabound.2018.07.004

[14] Bozhydarnyk V, Maksymovych O. Elastic equilibrium of an anisotropic half plane with periodic system of holes and cracks. Materials Science. 2001;**37**(6): 857-865. DOI: 10.1023/A:1015632905424

[15] Maksymovych O, Pasternak I, Sulym H, Kutsyk S. Doubly periodic cracks in the anisotropic medium with the account of contact of their faces. Acta Mechanica et Automatica. 2014;**8**(3): 160-164. DOI: 10.2478/ama-2014-0029

[16] Bozhydarnik VV, Maksymovych OV. Determination of the stressed state near edge cracks in a plate containing a hole of complex shape. Materials Science. 2010; **46**(1):16-26. DOI: 10.1007/s11003-010-9259-3

[17] Lekhnitskii SG. Anisotropic Plates. New York, London, Paris, Montreux, Tokyo, Melbourne: Gordon and Breach Science Publishers; 1987

Chapter 4

A New Computerized Boundary Element Model for Three-Temperature Nonlinear Generalized Thermoelastic Stresses in Anisotropic Circular Cylindrical Plate Structures

Mohamed Abdelsabour Fahmy

Abstract

In this chapter, we propose a new theory called nonlinear generalized thermoelasticity involving three temperatures. Because of strong nonlinearity of the proposed theory, therefore, it is much more difficult to develop analytical solution for solving problems related with the proposed theory. So, we propose a new computerized boundary element model for the solution of such problems and obtaining the three-temperature nonlinear generalized thermoelastic stresses in anisotropic circular cylindrical plate structures problems which are related with the proposed theory, where we used two-dimensional three temperature nonlinear radiative heat conduction equations coupled with electron, ion and phonon temperatures. The numerical results of the current study show the temperatures effects on the thermal stresses. Also, these numerical results demonstrate the validity and accuracy of our proposed model.

Keywords: boundary element model, three-temperature radiative heat conduction, nonlinear generalized thermoelasticity, thermal stresses, anisotropic circular cylindrical plate structures

1. Introduction

The spiral formed tube which has been used in water transmission pipelines [1, 2] is the most common structural application of a cylindrical shell. Spiral formed pipes were initially constructed by riveting together appropriately bent plates [3] until advances in welding technology allowed for efficient tandem arc welding [1]. Recently, increasing attention has been devoted to the study of spiral welded tubes due to its many applications in water, gas and oil pipelines under both low and high pressure [4] as well as for foundation piles and primary load-bearing members in Combi-walls [5]. Spiral welded tubes provide certain benefits over traditional longitudinal and butt-welded tubes. In particular, continuous or very long tubular members may be constructed efficiently from compact coils of metal strip, eliminating the need for costly transport of long tubular members. The coil material is usually manufactured to very tight tolerances which results in a tube with consistent wall thickness [6]. Further, they exhibit a superior fatigue performance compared to longitudinal seam welded tubes [7]. They also exhibit a comparable resistance to crack growth propagation in ductile materials [8]. However, spiral welded tubes are not suitable for offshore and deep-water applications, because their diameter and wall thickness are limited to nearly 3 m and 30 mm, respectively [9] which generally makes them unsuitable for offshore and deep-water applications [10].

In recent years, great attention has been directed towards the study of generalized thermoelastic interactions in anisotropic thermoelastic models due to its many applications in physics, geophysics, astronautics, aeronautics, earthquake engineering, military technologies, plasma, robotics, mining engineering, accelerators, nuclear reactors, nuclear plants, soil dynamics, automobile industries, high-energy particle accelerators and other science and engineering applications. The main notion of photons, which are particles of light energy, has been introduced by Albert Einstein in 1905. It is difficult to interpret why temperature depends on the specific heat of the crystalline solids. So, the original notion of phonons, which are particles of heat, has also introduced by Albert Einstein in 1907 to explain this phenomenon. Our three-temperature study is essential for a wide range of low-temperature applications, such as pool and basin heating, unglazed and uninsulated flat-plate organic collectors, cold storage warehouses, outdoor applications in extreme low temperatures, cryogenic gas processing plants and frozen food processing facilities. Also, our three-temperature study is very important high temperature applications such as turbine blades, piston engine valves, turbo charger components, microwave devices, laser diodes, RF power amplifiers, tubes of steam power plant, recuperators in the metallurgical and glass industries. The proposed boundary element method (BEM) can be easily implemented for solving nonlinear generalized thermoelasticity problems. Through the present paper, the three-temperature concept introduced for the first time in the field of nonlinear generalized thermoelasticity. Duhamel [11] and Neumann [12] developed the classical thermo-elasticity (CTE) theory and obtained the strain-temperature gradients equations in an elastic body, but their theory has the following two shortcomings: First, the heat conduction equation is predicting infinite speeds of propagation. Second, the heat conduction equation does not contain elastic terms. Biot [13] developed the classical coupled thermo-elasticity (CCTE) theory to overcome the first shortcoming in CTE. Then, several generalized theories based on a modified Fourier's law predict finite propagation speed of thermal waves such as extended thermo-elasticity (ETE) theory of Lord and Shulman (L-S) [14], temperature-rate-dependent thermo-elasticity (TRDTE) theory of Green and Lindsay (G-L) [15] and three linear generalized thermoelasticity models of Green and Naghdi (G-N) [16, 17], where Type I discusses the heat conduction theory based on Fourier's law, type II describes the thermoelasticity theory without energy dissipation (TEWOED), and type III discusses the thermoelasticity theory with energy dissipation (TEWED). Due to the computational difficulties, inherent in solving nonlinear generalized thermoelastic problems [18], for such problems, it is very difficult to obtain the analytical solution in a general case. Instead of analytical methods, many numerical methods were developed for solving such problems approximately including the finite difference method (FDM) [19, 20], discontinuous Galerkin method (DGM) [21], finite element method (FEM) [22, 23] and boundary element method (BEM) [24–26]. The boundary element method (BEM) has been performed successfully for solving various engineering, scientific and mathematical applications due to its simplicity, efficiency, and ease of implementation [27–46].

The main aim of the present chapter is to propose a new theory called nonlinear generalized thermoelasticity involving three-temperature. A new boundary element model was proposed for solving nonlinear generalized thermoelastic problems in anisotropic circular cylindrical plate structures which are associated with the proposed theory, where we used two-dimensional three-temperature (2D-3T) nonlinear time-dependent radiative heat conduction equations coupled with electron, ion and photon temperatures in the formulation of such problems. The numerical results are presented graphically to show the effects of electron, ion and photon temperatures on the thermal stress components. The validity and accuracy of our proposed BEM model were confirmed by comparing our BEM obtained results with the corresponding results of finite element method (FEM).

A brief summary of the chapter is as follows: Section 1 outlines the background and provides the readers with the necessary information to books and articles for a better understanding of mechanical behaviour of anisotropic circular cylindrical plate structures and their applications. Section 2 describes the formulation of the new theory and its related problems. Section 3 discusses the implementation of the new BEM for solving the three-temperature heat conduction equations, to obtain the temperature fields. Section 4 studies the development of new BEM and its implementation for solving the equilibrium equation based on the three-temperature fields. Section 5 presents the new numerical results that describe the temperatures effects on the thermal stresses generated in anisotropic circular cylindrical plate structures.

2. Formulation of the problem

We consider a cylindrical coordinate system (r, θ, z) for the circular cylindrical plate structure (**Figure 1**) within the region R which bounded by boundary S. Pressure distribution over the structure's entire surface has been shown in **Figure 2**. Geometry of meridional cross section of the considered structure has been shown in **Figure 3**, where $d\theta = \frac{1}{r}$.

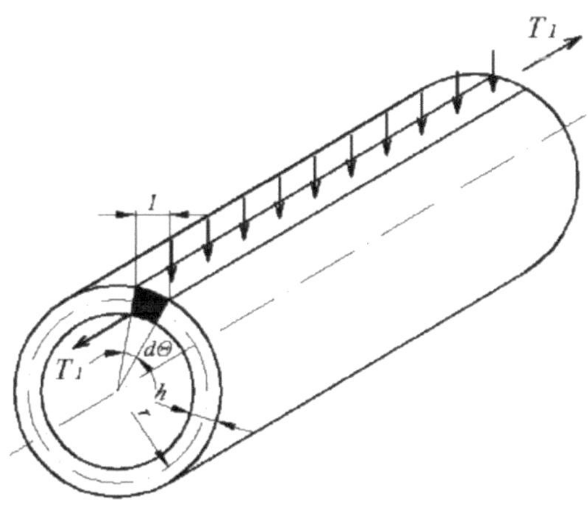

Figure 1.
Geometry of circular cylindrical plate structure.

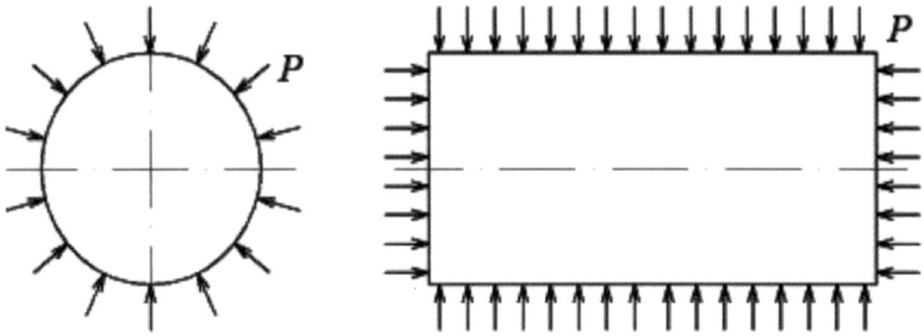

Figure 2.
Pressure distribution over the structure's entire surface.

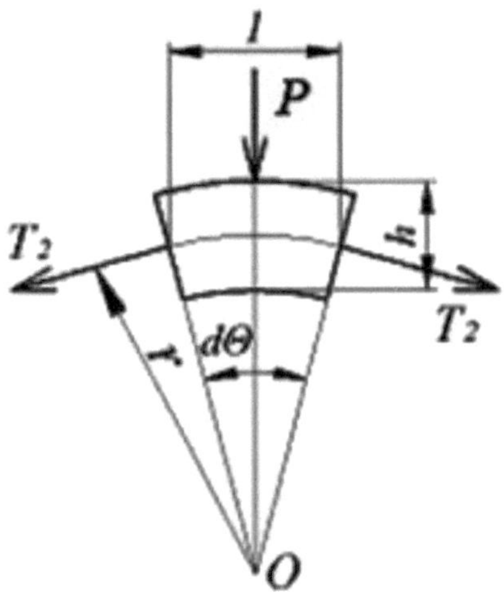

Figure 3.
Geometry of meridional cross section of considered structure.

The equilibrium equations for anisotropic plate structures can be written as follows

$$\sigma_{pj,j} = 0 \tag{1}$$

where

$$\sigma_{pj} = C_{pjkl} u_{k,l} - \beta_{pj} T_{\alpha}(r,\tau) \tag{2}$$

Three radiative heat conduction equations coupled with electron, ion and phonon temperatures can be written as follows

$$c_e \frac{\partial T_e(r,\tau)}{\partial \tau} - \frac{1}{\rho} \nabla[\mathbb{K}_e \nabla T_e(r,\tau)] = -\mathbb{W}_{ei}(T_e - T_i) - \mathbb{W}_{ep}(T_e - T_p) \tag{3}$$

$$c_i \frac{\partial T_i(r,\tau)}{\partial \tau} - \frac{1}{\rho} \nabla[\mathbb{K}_i \nabla T_i(r,\tau)] = \mathbb{W}_{ei}(T_e - T_i) \tag{4}$$

$$\frac{4}{\rho} c_p T_p^3 \frac{\partial T_p(r,\tau)}{\partial \tau} - \frac{1}{\rho} \nabla[\mathbb{K}_p \nabla T_p(r,\tau)] = \mathbb{W}_{ep}(T_e - T_p) \tag{5}$$

where (T_e, T_i, T_p), (c_e, c_i, c_p) and $(\mathbb{K}_e, \mathbb{K}_i, \mathbb{K}_p)$ are respectively temperatures, specific heat capacities and conductive coefficients of electron, ion and phonon.

The total temperature

$$T = T_e + T_i + T_p \tag{6}$$

3. BEM solution for three-temperature field

The nonlinear time-dependent two dimensions three temperature (2D-3T) radiative heat conduction Eqs. (3)–(5) coupled by electron, ion and phonon temperatures can be written as

$$\nabla\left[\left(\delta_{1j}\mathbb{K}_\alpha + \delta_{2j}\mathbb{K}_\alpha^*\right)\nabla T_\alpha(r,\tau)\right] - \overline{\mathbb{W}}(r,\tau) = c_\alpha \rho \delta_1 \delta_{1j} \frac{\partial T_\alpha(r,\tau)}{\partial \tau} \tag{7}$$

where

$$\overline{\mathbb{W}}(r,\tau) = \begin{cases} \rho\,\mathbb{W}_{ei}(T_e - T_i) + \rho\,\mathbb{W}_{er}\left(T_e - T_p\right) + \overline{\overline{\mathbb{W}}}, \alpha = e, \delta_1 = 1 \\ -\rho\mathbb{W}_{ei}(T_e - T_i) + \overline{\overline{\mathbb{W}}}, \quad \alpha = i, \delta_1 = 1 \\ -\rho\mathbb{W}_{er}\left(T_e - T_p\right) + \overline{\overline{\mathbb{W}}}, \quad \alpha = p, \delta_1 = \frac{4}{\rho}T_p^3 \end{cases} \tag{8}$$

$$\overline{\overline{\mathbb{W}}}(r,\tau) = -\delta_{2j}\mathbb{K}_\alpha \dot{T}_{\alpha,ab} + \beta_{ab}T_{\alpha 0}\left[\mathring{A}\delta_{1j}\dot{u}_{a,b} + \left(\tau_0 + \delta_{2j}\right)\ddot{u}_{a,b}\right] + \rho c_\alpha\left[\left(\tau_0 + \delta_{1j}\tau_2 + \delta_{2j}\right)\ddot{T}_\alpha\right] - \frac{T_{\alpha,a}}{r}, \quad \frac{T_{\alpha,a}}{r} = T_{\alpha,aa} \tag{9}$$

$$\mathbb{W}_{ei} = \rho\mathbb{A}_{ei}T_e^{-2/3}, \mathbb{W}_{er} = \rho\mathbb{A}_{er}T_e^{-1/2}, \mathbb{K}_\alpha = \mathbb{A}_\alpha T_\alpha^{5/2}, \alpha = e, i, \mathbb{K}_p = \mathbb{A}_p T_p^{3+\mathbb{B}} \tag{10}$$

The total energy can be written as follows

$$P = P_e + P_i + P_p, P_e = c_e T_e, P_i = c_i T_i, P_p = \frac{1}{\rho}c_p T_p^4 \tag{11}$$

By applying the following conditions

$$T_\alpha(x,y,0) = T_\alpha^0(x,y) = g_1(x,\tau) \tag{12}$$

$$\mathbb{K}_\alpha \frac{\partial T_\alpha}{\partial n}\bigg|_{\Gamma_1} = 0, \alpha = e, i, T_r|_{\Gamma_1} = g_2(x,\tau) \tag{13}$$

$$\mathbb{K}_\alpha \frac{\partial T_\alpha}{\partial n}\bigg|_{\Gamma_2} = 0, \alpha = e, i, p \tag{14}$$

By using the fundamental solution that satisfies the following Eq. [46]

$$D\nabla^2 T_\alpha + \frac{\partial T_\alpha^*}{\partial n} = -\delta(r - p_i)\delta(\tau - r) \tag{15}$$

where $D = \frac{\mathbb{K}_\alpha}{\rho c}$ and p_i are singular points.
The corresponding dual reciprocity boundary integral equation can be written as [46]

$$CT_\alpha = \frac{D}{\mathbb{K}_\alpha}\int_O^\tau\int_S \left[T_\alpha q^* - T_\alpha^* q\right] dS d\tau + \frac{D}{\mathbb{K}_\alpha}\int_O^\tau\int_R bT_\alpha^* dR d\tau + \int_R T_\alpha^i T_\alpha^* \big|_{\tau=0} dR \tag{16}$$

which can be expressed as

$$CT_\alpha = \int_S [T_\alpha q^* - T_\alpha^* q] dS - \int_R \frac{\mathbb{K}_\alpha}{D} \frac{\partial T_\alpha^*}{\partial \tau} T_\alpha dR \tag{17}$$

In order to transform the domain integral into the boundary, we assume that

$$\frac{\partial T_\alpha}{\partial \tau} \cong \sum_{j=1}^{N} f^j(r)^j a^j(\tau) \tag{18}$$

where $f^j(r)$ and $a^j(\tau)$ are known functions and unknown coefficients, respectively.

We assume that $\hat{T}_\alpha^j$ is a solution of

$$\nabla^2 \hat{T}_\alpha^j = f^j \tag{19}$$

Thus, from (17) we can write the following boundary integral equation

$$CT_\alpha = \int_S [T_\alpha q^* - T_\alpha^* q] dS + \sum_{j=1}^{N} a^j(\tau) D^{-1} \left(C\hat{T}_\alpha^j - \int_S [\hat{T}_\alpha^j q^* - \hat{q}^j T_\alpha^*] dS \right) \tag{20}$$

where

$$\hat{q}^j = -\mathbb{K}_\alpha \frac{\partial \hat{T}_\alpha^j}{\partial n} \tag{21}$$

$$a^j(\tau) = \sum_{i=1}^{N} f_{ji}^{-1} \frac{\partial T(r_i, \tau)}{\partial \tau} \tag{22}$$

$$\{F\}_{ji} = f^j(r_i) \tag{23}$$

By using (20) and (22), we obtain

$$C\dot{T}_\alpha + HT_\alpha = GQ \tag{24}$$

where

$$C = -\left[H\hat{T}_\alpha - G\hat{Q} \right] F^{-1} D^{-1} \tag{25}$$

and

$$\{\hat{T}\}_{ij} = \hat{T}^j(x_i) \tag{26}$$

$$\left\{\hat{Q}\right\}_{ij} = \hat{q}^j(x_i) \tag{27}$$

For solving (24) numerically, the functions q, T_α and its derivative with time can be written as

$$q = (1 - \Theta) q^m + \Theta\, q^{m+1}, 0 \leq \Theta \leq 1 \tag{28}$$

$$T_\alpha = (1-\Theta)T_\alpha^m + \theta T_\alpha^{m+1}, 0 \le \Theta = \frac{\tau - \tau^m}{\tau^{m+1} - \tau^m} \le 1 \tag{29}$$

$$\dot{T}_\alpha = \frac{dT_\alpha}{d\Theta}\frac{d\Theta}{d\tau} = \frac{T_\alpha^{m+1} - T_\alpha^m}{\tau^{m+1} - \tau^m} = \frac{T_\alpha^{m+1} - T_\alpha^m}{\Delta\tau^m} \tag{30}$$

By substituting from Eqs. (28)–(30) into Eq. (24), we obtain

$$\left(\frac{C}{\Delta\tau^m} + \Theta H\right)T_\alpha^{m+1} - \Theta G Q^{m+1} = \left(\frac{C}{\Delta\tau^m} - (1-\Theta)H\right)T_\alpha^m + (1-\Theta)GQ^m \tag{31}$$

By applying the initial and boundary conditions, we obtain

$$\mathbb{a}\mathrm{X} = \mathbb{b} \tag{32}$$

This system yields the temperature in terms of the displacement field.

4. BEM solution for displacement field

The equilibrium Eqs. (1) for anisotropic plate structures can be written as follows [47]

$$C_{ijkl}\frac{d^4w}{dx^4} - T\frac{d^2w}{dx^2} = p + \frac{T_1}{r} \tag{33}$$

where

$$T_2 = -\frac{pr}{2} \tag{34}$$

$$T_1 = C_{ijkl}T_2 - C_{ijkl}h\frac{w}{r} \tag{35}$$

By using (34) and (35), we can write (33) in the following form

$$C_{ijkl}\frac{d^4w}{dx^4} + \frac{pr}{2}\frac{d^2w}{dx^2} + \frac{C_{ijkl}h}{r^2}w = p\left(1 - \frac{C_{ijkl}}{2}\right) \tag{36}$$

where

$$\mathrm{A} = \frac{C_{ijkl}h}{r^2} \tag{37}$$

$$\mathrm{B} = p\left(1 - \frac{C_{ijkl}}{2}\right) \tag{38}$$

By using Eqs. (37) and (38), we can write (36) as follows

$$C_{ijkl}\frac{d^4w}{dx^4} - T\frac{d^2w}{dx^2} + \mathrm{A}w = \mathrm{B} \tag{39}$$

where

$$\beta = \frac{T_2}{2\sqrt{C_{ijkl}k}}, 0 < \beta^2 < 1 \tag{40}$$

The general solution of (39) can be obtained as

$$w(x) = C_1 ch\delta x \cos\gamma x + C_2 ch\delta x \sin\gamma x + C_3 sh\delta x \cos\gamma x + C_4 sh\delta x \sin\gamma x + w_{\text{part}} \tag{41}$$

where

$$\delta = \alpha\sqrt{1+\beta};\gamma = \alpha\sqrt{1-\beta};\alpha = \sqrt[4]{\frac{k}{4C_{ijkl}}},\beta = \frac{T_2}{2\sqrt{C_{ijkl}k}} \tag{42}$$

and the particular solution can be determined as p = constant as follows

$$w_{\text{part}} = \frac{pr^2}{C_{ijkl}h}\left(1 - \frac{C_{ijkl}}{2}\right) \tag{43}$$

Thus, Eq. (41) can be written as

$$w(x) = \frac{pr^2}{C_{ijkl}h}\left(1 - \frac{C_{ijkl}}{2}\right) + C_1 ch\delta x \cos\gamma x + C_4 sh\delta x \sin\gamma x \tag{44}$$

By implementing the following boundary conditions.

$$\text{at } x = \pm\frac{l}{2}\frac{dw}{dx} = 0 \tag{45}$$

$$\text{at } x = \frac{l}{2}w = \frac{2pr^2}{C_{ijkl}h}\frac{d^3w}{dx^3} \tag{46}$$

we can write the unknown C_1 and C_4 as follows

$$C_1 = -\frac{2pr^2}{C_{ijkl}h}\left(1 - \frac{C_{ijkl}}{2}\right)\frac{u_1 chu_1 \sin u_2 + u_2 shu_1 \cos u_2}{u_2 sh2u_1 + u_1 \sin 2u_2}\varepsilon_1 \tag{47}$$

$$C_4 = -\frac{2pr^2}{C_{ijkl}h}\left(1 - \frac{C_{ijkl}}{2}\right)\frac{u_2 chu_1 \sin u_2 - u_1 shu_1 \cos u_2}{u_2 sh2u_1 + u_1 \sin 2u_2}\varepsilon_1 \tag{48}$$

where

$$\varepsilon_1 = \frac{1}{1 + \frac{lh}{A}A_1(u_1, u_2)} \tag{49}$$

$$A_1(u_1, u_2) = \sqrt{1-\beta^2}\frac{ch2u_1 - \cos 2u_2}{u_2 sh2u_1 + u_1 \sin 2u_2} \tag{50}$$

$$u_1 = \frac{\delta l}{2} = u\sqrt{1+\beta}, u_2 = \frac{\gamma l}{2} = u\sqrt{1+\beta}, u = 0.6425\frac{1}{\sqrt{rh}} \tag{51}$$

If we neglected the longitudinal forces influence on the bending of the circular cylindrical shell, we can write (39) in the following form

$$C_{ijkl}w^{IV} + kw = q \tag{52}$$

Now, the approximate solution has been reduced for solving problem of bending single span beam with the following compliance

$$k_{II} = \frac{2r^2}{C_{ijkl}A} \tag{53}$$

The deflection of the considered shell in the cross section and reference section, respectively, is as follows

$$w(0) = \frac{pr^2}{C_{ijkl}h}\left(1 - \frac{v}{2}\right)\left[1 - \frac{\varnothing_1(u)}{1 + B_1}\right] \tag{54}$$

$$w\left(\frac{l}{2}\right) = \frac{pr^2}{C_{ijkl}h}\left(1 - \frac{v}{2}\right)\frac{B_1}{1 + B_1} \tag{55}$$

Also, the bending moment in the cross section and reference section, respectively, is as follows

$$M_1(0) = -\frac{pl^2}{24}\left(1 - \frac{v}{2}\right)\left[1 - \frac{\chi_1(u)}{1 + B_1}\right] \tag{56}$$

$$M_1\left(\frac{l}{2}\right) = \frac{pl^2}{12}\left(1 - \frac{v}{2}\right)\frac{\chi_2(u)}{1 + B_1} \tag{57}$$

The Cauchy model with two-bed scheme can be described as follows

$$Dv^{IY}(x) + \frac{pr}{2}v''(x) + \frac{C_{ijkl}h}{r^2}v(x) = p\left(1 - \frac{\mu}{2}\right) \tag{58}$$

$$v(0); \varphi(0) = v'(0) \tag{59}$$

$$M(0) = -Dv''(0) - Tv(0) \tag{60}$$

$$Q(0) = -Dv'''(0) - Tv'(0) \tag{61}$$

where the characteristic equation of (58) can be defined as

$$C_{ijkl}k^4 + \frac{pr}{2}k^2 + \frac{C_{ijkl}h}{r^2} = 0, k^2 = t \tag{62}$$

$$C_{ijkl}t^2 + \frac{pr}{2}t + \frac{C_{ijkl}h}{r^2} = 0 \tag{63}$$

which roots

$$k_{1,2,3,4} = \pm\sqrt{\frac{-\frac{pr}{2} \pm \sqrt{\left(\frac{pr}{2}\right)^2 - 4{C_{ijkl}}^2\frac{h}{r^2}}}{2C_{ijkl}}} \tag{64}$$

$$t_{1,2} = \frac{-\frac{pr}{2} \pm \sqrt{\left(\frac{pr}{2}\right)^2 - 4{C_{ijkl}}^2\frac{h}{r^2}}}{2C_{ijkl}} \tag{65}$$

The systems (32) and (58) can be solved by using the algorithm of Fahmy [35] to obtain the three temperatures and displacements components. Then we can compute thermal stresses distributions along radial distance r. we refer the reader to recent references [48–51] for details of boundary element technique.

5. Numerical results and discussion

The BEM that has been used in the current chapter can be applicable to a wide variety *of plate structures problems* associated with the proposed theory of three temperatures nonlinear generalized thermoelasticity. In order to evaluate temperatures effects on the thermal stresses, the numerical results are carried out and depicted graphically for electron, ion and phonon temperatures.

Figure 4 shows the distributions of the three temperatures T_e, T_i, T_p and total temperature T $\left(T = T_e + T_i + T_p\right)$ along the radial distance r. It was shown from this figure that the three temperatures are different and they may have great effects on the connected fields.

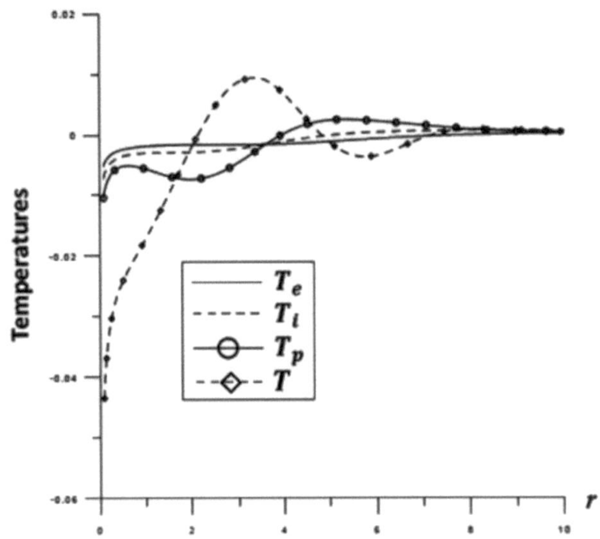

Figure 4.
Variation of the temperatures T_e, T_i, T_p and T along the radial distance r.

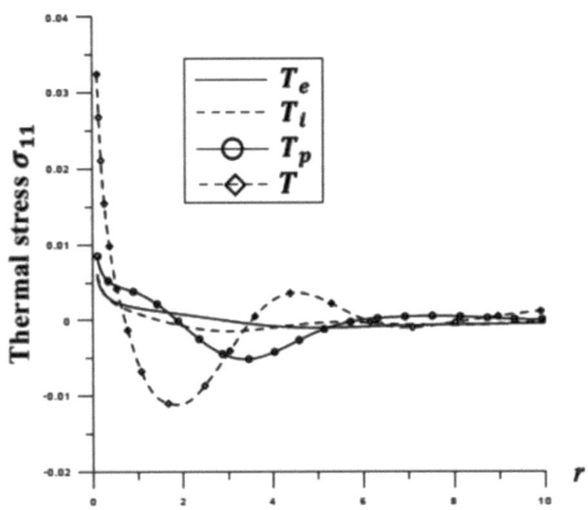

Figure 5.
Variation of the thermal stress σ_{11} with radial distance r.

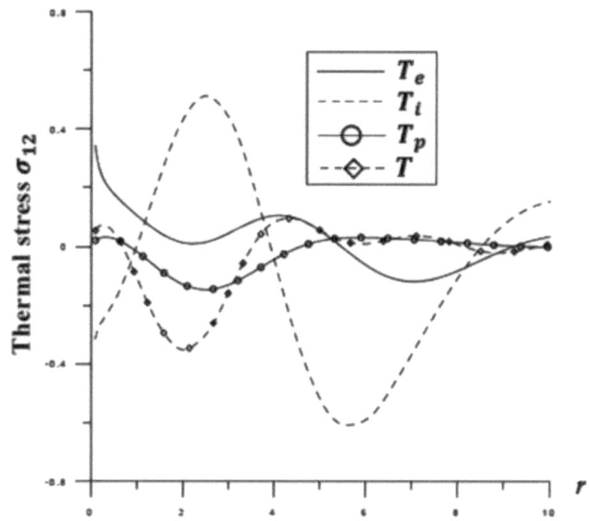

Figure 6.
Variation of the thermal stress σ_{12} with radial distance r.

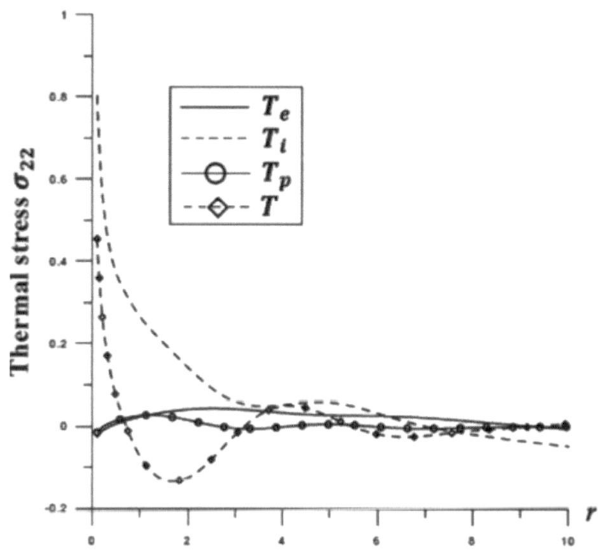

Figure 7.
Variation of the thermal stress σ_{22} with radial distance r.

Figures 5–7 show the distributions of the thermal stresses σ_{11}, σ_{12} and σ_{22} respectively, with the radial distance r for the three temperatures T_e, T_i, T_p and total temperature. It was noticed from these figures that the three temperatures have great effects on the thermal stresses.

Figure 8 shows the distributions of the thermal stresses $\sigma_{11}, \sigma_{12}, \sigma_{22}$ and total temperature T with the radial distance r for BEM results and finite element method (FEM) results of COMSOL Multiphysics software version 5.4 to demonstrate the validity and accuracy of our proposed model based on replacing heat conduction with three-temperature heat conduction.

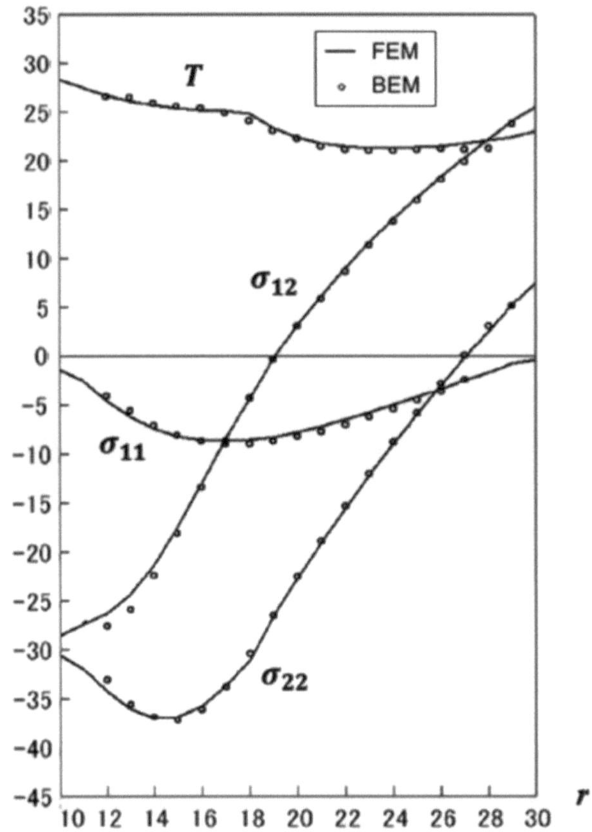

Figure 8.
Thermal stresses and total temperature variations with r.

6. Conclusion

The main objective of this chapter is to propose a new theory called nonlinear generalized thermoelasticity involving three-temperature and new BEM model for the solution of problems which are associated with the proposed nonlinear theory, where we used the three-temperature radiative heat conduction equations coupled with electron, ion and phonon temperatures to describe the thermal stresses in anisotropic circular cylindrical plate structures. It can be concluded from numerical results of our proposed model that the generalized theories of thermoelasticity can be connected with the three-temperature radiative heat conduction to describe the deformation of anisotropic circular cylindrical plate structures. The validity and accuracy of the proposed model was examined and confirmed by comparing the obtained results with those known previously. Because there are no available data to confirm the validity and accuracy of our results, we replace the three-temperature radiative heat conduction results with one-temperature heat conduction results as a special case from results of our current general model for circular cylindrical plate structures. In the special case under consideration, the results obtained with the BEM have been compared graphically with the FEM results of COMSOL Multiphysics software version 5.4. Excellent agreement is obtained between BEM results and FEM results. Understanding the behaviour of the three-temperature thermal stresses in anisotropic circular cylindrical plate structures should be a key for extending the application of these behaviors to a wide range of structures. The numerical results for our general model which is associated with our proposed theory may provide interesting information for computer scientists and engineers, geotechnical and geothermal engineers, researchers who will industrialize the thermoelastic devices using additive manufacturing and the materials designers and developers, etc.

Nomenclature

β_{ij}	stress-temperature coefficients
δ_{ij}	Kronecker delta $(i, j = 1, 2)$
ε_{ij}	strain tensor
θ	thermodynamic temperature
λ	tractions
μ_0	magnetic permeability
ϑ_0	viscoelastic relaxation time
ϖ	weights of control points
ρ	density
σ_{ij}	force stress tensor
c	specific heat capacity
C_{ijkl}	constant elastic moduli
$\mathbb{e}_{lij}$	piezoelectric tensor
F_i	mass force vector
$\mathbb{K}_\alpha$	conductive coefficients
M_1	bending moment
P	total energy of unit mass
T_α	temperature functions
u_i	displacement vector
$w(x)$	general solution
$\mathbb{W}_{ei}$	electron-ion energy coefficient
$\mathbb{W}_{ep}$	electron-photon energy coefficient

Author details

Mohamed Abdelsabour Fahmy
Faculty of Computers and Informatics, Suez Canal University, Ismailia, Egypt

*Address all correspondence to: mohamed_fahmy@ci.suez.edu.eg

References

[1] Hanada M, Takeda H, Fukushima H, Koizumi I, Noguchi Y. Development of highspeed submerged arc welding in spiral pipe mill. Transactions of the Iron and Steel Institute of Japan. 1986;**26**: 433-438

[2] Burstall T. Bulk Water Pipelines. London, United Kingdom: Thomas Telford Ltd.; 1997

[3] Hardenbergh WA. Water Supply and Purification. 2nd ed. Scranton, PA: International Textbook Company; 1945

[4] Liu H. Pipeline engineering. Florida, United States: CRC Press; 2003

[5] CUR. Handbook Quay Walls– Publication N. 211 E. Gouda, Holland: Centre for Civil Engineering Research and Codes (CUR); 2005

[6] Lancaster J. Handbook of Structural Welding: Processes, Materials and Methodsused in the Welding of Major Structures, Pipelines and Process Plants. Cambridge, United Kingdom: Woodhead Publishing; 1997

[7] Radaj D. Design and Analysis of Fatigue Resistant Welded Structures. Cambridge, United Kingdom: Woodhead Publishing; 1990

[8] Andrews RM, Pistone V. European pipeline research group studies on ductilecrack propagation in gas transmission pipelines. European Structural Integrity Society. 2002;**30**: 377-384

[9] Mittal A. Spirally Welded Steel Pipes. The Netherlands: Corporate Brochure, Projects-Europe, Foundation Solutions; 2011

[10] Gerwick BC. Construction of Marine and Offshore Structures. 3rd ed. Florida, United States: CRC Press; 2007

[11] Duhamel J. Some memoire sur les phenomenes thermo-mechanique. Journal de l'École Polytechnique. 1837; **15**:1-57

[12] Neumann F. Vorlesungen Uber die theorie der elasticitat. Brestau: Meyer; 1885

[13] Biot M. Thermoelasticity and irreversible thermo-dynamics. Journal of Applied Physics. 1956;**27**:249-253

[14] Lord HW, Shulman Y. A generalized dynamical theory of thermoelasticity. Journal of the Mechanics and Physics of Solids. 1967; **15**:299-309

[15] Green AE, Lindsay KA. Thermoelasticity. Journal of Elasticity. 1972;**2**:1-7

[16] Green AE, Naghdi PM. On undamped heat waves in an elastic solid. Journal of Thermal Stresses. 1992;**15**: 253-264

[17] Green AE, Naghdi PM. Thermoelasticity without energy dissipation. Journal of Elasticity. 1993; **31**:189-208

[18] Jafarian A, Ghaderi P, Golmankhaneh AK, Baleanu D. Analytic solution for a nonlinear problem of magneto-Thermoelasticity. Reports on Mathematical Physics. 2013;**71**:399-411. DOI: 10.1016/S0034-4877(13)60039-7

[19] Abd-El-Salam MR, Abd-Alla AM, Hosham HA. A numerical solution of magneto-thermoelastic problem in non-homogeneous isotropic cylinder by the finite-difference method. Applied Mathematical Modelling. 2007;**31**: 1662-1670. DOI: 10.1016/j.apm.2006.05.009

[20] Abd-Alla AM, El-Naggar AM, Fahmy MA. Magneto-thermoelastic problem in non-homogeneous isotropic cylinder. Heat and Mass Transfer. 2003; **39**:625-629

[21] Hu Q, Zhao L. Domain decomposition preconditioners for the system generated by discontinuous Galerkin discretization of 2D-3T heat conduction equations. Communications in Computational Physics. 2017;**22**: 1069-1100

[22] Abbas IA, Abd-alla AN, Othman MIA. Generalized magneto-thermoelasticity in a fiber-reinforced anisotropic half-space. International Journal of Thermophysics. 2011;**32**: 1071-1085. DOI: 10.1007/s10765-011-0957-3

[23] Abbas IA. Generalized magneto-thermoelasticity in a nonhomogeneous isotropic hollow cylinder using the finite element method. Archive of Applied Mechanics. 2009;**79**:41-50. DOI: 10.1007/s00419-008-0206-9

[24] Abd-Alla AM, Fahmy MA, El-Shahat TM. Magneto-thermo-elastic problem of a rotating non-homogeneous anisotropic solid cylinder. Archive of Applied Mechanics. 2008;**78**:135-148

[25] Fahmy MA. A time-stepping DRBEM for magneto-thermo-viscoelastic interactions in a rotating nonhomogeneous anisotropic solid. International Journal of Applied Mechanics. 2011;**3**:1-24

[26] Fahmy MA. A time-stepping DRBEM for the transient magneto-thermo-visco-elastic stresses in a rotating non-homogeneous anisotropic solid. Engineering Analysis with Boundary Elements. 2012;**36**: 335-345

[27] Fahmy MA. Boundary element modeling and simulation of biothermomechanical behavior in anisotropic laser-induced tissue hyperthermia. Engineering Analysis with Boundary Elements. 2019;**101**: 156-164

[28] Fahmy MA. Boundary element algorithm for modeling and simulation of dual-phase lag bioheat transfer and biomechanics of anisotropic soft tissues. International Journal of Applied Mechanics. 2018;**10**:1850108

[29] Fahmy MAA. New computerized boundary element algorithm for cancer modeling of cardiac anisotropy on the ECG simulation. Asian Journal of Research in Computer Science. 2018;**2**: 1-10

[30] Fahmy MA. Transient magneto-thermoviscoelastic plane waves in a non-homogeneous anisotropic thick strip subjected to a moving heat source. Applied Mathematical Modelling. 2012; **36**:4565-4578

[31] Fahmy MA. Numerical modeling of transient magneto-thermo-viscoelastic waves in a rotating nonhomogeneous anisotropic solid under initial stress. International Journal of Modeling, Simulation and Scientific Computing. 2012;**3**:1250002

[32] Fahmy MA. The effect of rotation and inhomogeneity on the transient magneto-thermoviscoelastic stresses in an anisotropic solid. ASME Journal of Applied Mechanics. 2012;**79**:1015

[33] Fahmy MA. Transient magneto-thermo-viscoelastic stresses in a rotating nonhomogeneous anisotropic solid with and without a moving heat source. Journal of Engineering Physics and Thermophysics. 2012;**85**: 950-958

[34] Fahmy MA. Transient magneto-thermo-elastic stresses in an anisotropic viscoelastic solid with and without

moving heat source. Numerical Heat Transfer, Part A: Applications. 2012;**61**: 547-564

[35] Fahmy MA. Implicit-explicit time integration DRBEM for generalized magneto-thermoelasticity problems of rotating anisotropic viscoelastic functionally graded solids. Engineering Analysis with Boundary Elements. 2013; **37**:107-115

[36] Fahmy MA. Generalized magneto-thermo-viscoelastic problems of rotating functionally graded anisotropic plates by the dual reciprocity boundary element method. Journal of Thermal Stresses. 2013;**36**:1-20

[37] Fahmy MA. A three-dimensional generalized magneto-thermo-viscoelastic problem of a rotating functionally graded anisotropic solids with and without energy dissipation. Numerical Heat Transfer, Part A: Applications. 2013;**63**:713-733

[38] Fahmy MA. A computerized DRBEM model for generalized magneto-thermo-visco-elastic stress waves in functionally graded anisotropic thin film/substrate structures. Latin American Journal of Solids and Struc tures. 2014;**11**:386-409

[39] Fahmy MA. Computerized Boundary Element Solutions for Thermoelastic Problems: Applications to Functionally Graded Anisotropic Structures. Saarbrücken, Germany: LAP Lambert Academic Publishing; 2017

[40] Fahmy MA. Boundary Element Computation of Shape Sensitivity and Optimization: Applications to Functionally Graded Anisotropic Structures. Saarbrücken, Germany: LAP Lambert Academic Publishing; 2017

[41] Fahmy MA. Shape design sensitivity and optimization for two-temperature generalized magneto-thermoelastic problems using time-domain DRBEM. Journal of Thermal Stresses. 2018;**41**: 119-138

[42] Fahmy MA. Shape design sensitivity and optimization of anisotropic functionally graded smart structures using bicubic B-splines DRBEM. Engineering Analysis with Boundary Elements. 2018;**87**:27-35

[43] Fahmy MA. Modeling and optimization of anisotropic viscoelastic porous structures using CQBEM and moving asymptotes algorithm. Arabian Journal for Science and Engineering. 2019;**44**:1671-1684

[44] Fahmy MA. A new LRBFCM-GBEM modeling algorithm for general solution of time fractional order dual phase lag bioheat transfer problems in functionally graded tissues. Numerical Heat Transfer, Part A: Applications. 2019;**75**:616-626

[45] Fahmy MA. Design optimization for a simulation of rotating anisotropic viscoelastic porous structures using time-domain OQBEM. Mathematics and Computers in Simulation. 2019;**66**: 193-205

[46] Fahmy MA. A new boundary element strategy for modeling and simulation of three temperatures nonlinear generalized micropolar-magneto-thermoelastic wave propagation problems in FGA structures. Engineering Analysis with Boundary Elements. 2019;**108**:192-200

[47] Aniskin A, Orobey VF, Kolomiets LV, Lymarenko AM. Calculation of displacements and stresses in cylindrical shells by the boundary elements method. Technical Journal. 2018;**12**:196-203

[48] Fahmy MA. DRBEM sensitivity analysis and shape optimization of rotating magneto-thermo-viscoelastic FGA structures using DRBEM-GSS and DRBEM-NGGP algorithms. In:

Advances in Mathematics and Computer Science. Vol. 1. London, United Kingdom: Book Publisher International; 2019. pp. 83-104

[49] Fahmy MA. The effect of anisotropy on the structure optimization using BEM-GSS and BEM-NGGP algorithms. In: Advances in Mathematics and Computer Science. Vol. 1. London, United Kingdom: Book Publisher International; 2019. pp. 137-156

[50] Fahmy MA, Al-Harbi SM, Al-Harbi BH, Sibih AM. A computerized boundary element algorithm for modeling and optimization of complex magneto-thermoelastic problems in MFGA structures. In: Advances in Applied Science and Technology. Vol. 2. London, United Kingdom: Book Publisher International; 2019. pp. 139-152

[51] Fahmy MA. Boundary element model for nonlinear fractional-order heat transfer in magneto-thermoelastic FGA structures involving three temperatures. In: Mechanics of Functionally Graded Materials and Structures. London, United Kingdom: IntechOpen; 2019

Chapter 5

Distinctive Characteristics of Cosserat Plate Free Vibrations

Lev Steinberg and Roman Kvasov

Abstract

In this chapter, we present the theoretical analysis of the distinctive characteristics of Cosserat plate vibrations. This analysis is based on the dynamic model of the Cosserat plates, which we developed as an extension of the Reissner plate theory. Primarily, we describe the validation of the model, which is based on the comparison with three-dimensional exact solutions. We present the results of the computer simulations, which allow us to identify different characteristics of the plate vibrations. Particularly, we illustrate and discuss the detection and the classification of the additional high resonance frequencies of a plate depending on the shape and orientation of microelements incorporated into the Cosserat plates.

Keywords: variational principle, Cosserat plate vibrations, frequencies of micro-vibrations

1. Introduction

The theory of asymmetric elasticity introduced in 1909 by the Cosserat brothers [1] gave rise to a variety of Cosserat plate theories. In 1960s, Green and Naghdi specialized their general theory of Cosserat surface to obtain the linear Cosserat plate [2], while independently Eringen proposed a complete theory of plates in the framework of Cosserat elasticity [3]. Numerous plate theories were formulated afterwards; for the review of the latest developments in the area of Cosserat plates we recommend to turn to [4].

The first theory of Cosserat plates based on the Reissner plate theory was developed in [5] and its finite element modeling is provided in [6]. The parametric theory of Cosserat plate, presented by the authors in [7], includes some additional assumptions leading to the introduction of the splitting parameter. This provided the highest level of approximation to the original three-dimensional problem. The theory provides the equilibrium equations and constitutive relations, and the optimal value of the minimization of the elastic energy of the Cosserat plate. The paper [7] also provides the analytical solutions of the presented plate theory and the three-dimensional Cosserat elasticity for simply supported rectangular plate. The comparison of these solutions showed that the precision of the developed Cosserat plate theory is similar to the precision of the classical plate theory developed by Reissner [8, 9].

The numerical modeling of bending of simply supported rectangular plates is given in [10]. We developed the Cosserat plate field equations and a rigorous formula for the optimal value of the splitting parameter. The solution of the

Cosserat plate was shown to converge to the Reissner plate as the elastic asymmetric parameters tend to zero. The Cosserat plate theory demonstrates the agreement with the size effect, confirming that the plates of smaller thickness are more rigid than is expected from the Reissner model. The modeling of Cosserat plates with simply supported rectangular holes is also provided. The finite element analysis of the perforated Cosserat plates is given in [11].

The extension of the static model of Cosserat elastic plates to the dynamic problems is presented in [12]. The computations predict a new kind of natural frequencies associated with the material microstructure and were shown to be compatible with the size effect principle reported in [10] for the Cosserat plate bending.

This chapter represents an extension of the paper [12] for different shapes and orientations of micro-elements incorporated into the Cosserat plates. It is based on the generalized variational principle for elastodynamics and includes a non-diagonal rotatory inertia tensor. The numerical computations of the plate free vibrations showed the existence of some additional high frequencies of micro-vibrations depending on the orientation of micro-elements. The comparison with three-dimensional Cosserat elastodynamics shows a high agreement with the exact values of the eigenvalue frequencies.

2. Cosserat linear elastodynamics

2.1 Fundamental equations

The Cosserat linear elasticity balance laws are

$$\sigma_{ji,j} = \frac{\partial p_i}{\partial t}, \tag{1}$$

$$\varepsilon_{ijk}\sigma_{jk} + \mu_{ji,j} = \frac{\partial q_i}{\partial t}, \tag{2}$$

where the σ_{ji} is the stress tensor, μ_{ji} the couple stress tensor, $p_i = \rho\frac{\partial u_i}{\partial t}$ and $q_i = J_{ji}\frac{\partial \phi_j}{\partial t}$ are the linear and angular momenta, ρ and J_{ji} are the material density and the rotatory inertia characteristics, ε_{ijk} is the Levi-Civita tensor.

We will also consider the constitutive equations as in [13]:

$$\sigma_{ji} = (\mu + \alpha)\gamma_{ji} + (\mu - \alpha)\gamma_{ij} + \lambda\gamma_{kk}\delta_{ij}, \tag{3}$$

$$\mu_{ji} = (\gamma + \varepsilon)\chi_{ji} + (\gamma - \varepsilon)\chi_{ij} + \beta\chi_{kk}\delta_{ij}, \tag{4}$$

and the kinematic relations in the form

$$\gamma_{ji} = u_{i,j} + \varepsilon_{ijk}\phi_k \text{ and } \chi_{ji} = \phi_{i,j}, \tag{5}$$

Here u_i and ϕ_i represent the displacement and microrotation vectors, γ_{ji} and χ_{ji} represent the strain and bend-twist tensors, μ, λ are the Lamé parameters and α, β, γ, ε are the Cosserat elasticity parameters.

The constitutive Eqs. (3)–(4) can be written in the reverse form [5].

$$\gamma_{ji} = (\mu' + \alpha')\sigma_{ji} + (\mu' - \alpha')\sigma_{ij} + \lambda'\sigma_{kk}, \tag{6}$$

$$\chi_{ji} = (\gamma' + \varepsilon')\mu_{ji} + (\gamma' - \varepsilon')\mu_{ij} + \beta'\mu_{kk}, \tag{7}$$

where $\mu' = \frac{1}{4\mu}$, $\alpha' = \frac{1}{4\alpha}$, $\gamma' = \frac{1}{4\gamma}$, $\varepsilon = \frac{1}{4\varepsilon}$, $\lambda' = \frac{-\lambda}{6\mu\left(\lambda + \frac{2\mu}{3}\right)}$ and $\beta' = \frac{-\beta}{6\mu\left(\beta + \frac{2\gamma}{3}\right)}$.

We will consider the boundary conditions given in [12].

$$u_i = u_i^0,\ \phi_i = \phi_i^0,\ \text{on}\ \mathcal{G}_1^t = \partial B_0 \backslash \partial B_\sigma \times [t_0, t], \tag{8}$$

$$\sigma_{ji} n_i = \sigma_j^0,\ \mu_{ji} n_i = \mu_j^0\ \text{on}\ \mathcal{G}_2^t = \partial B_\sigma \times [t_0, t], \tag{9}$$

and initial conditions

$$u_i(x, 0) = U_i^0,\ \phi_i(x, 0) = \Phi_i^0,\ \text{in}\, B_0, \tag{10}$$

$$\dot{u}_i(x, 0) = \dot{U}_i^0,\ \dot{\phi}_i(x, 0) = \dot{\Phi}_i^0,\ \text{in}\, B_0, \tag{11}$$

where u_i^0 and ϕ_i^0 are prescribed on $\mathcal{G}_1$, σ_j^0 and μ_j^0 on $\mathcal{G}_2$, and n_i is the unit vector normal to the boundary ∂B_0 of the elastic body B_0.

2.2 Cosserat elastic energy

The strain stored energy U_C of the body B_0 is defined by the integral [13]:

$$U_C = \int_{B_0} \mathrm{W}\{\gamma, \chi\} dv, \tag{12}$$

where

$$\begin{aligned} \mathrm{W}\{\gamma, \chi\} = &\frac{\mu + \alpha}{2}\gamma_{ij}\gamma_{ij} + \frac{\mu - \alpha}{2}\gamma_{ij}\gamma_{ji} + \frac{\lambda}{2}\gamma_{kk}\gamma_{nn} \\ &+ \frac{\gamma + \varepsilon}{2}\chi_{ij}\chi_{ij} + \frac{\gamma - \varepsilon}{2}\chi_{ij}\chi_{ji} + \frac{\beta}{2}\chi_{kk}\chi_{nn}, \end{aligned} \tag{13}$$

is non-negative. The relations Eqs. (3)–(4) can be written in the form [12]:

$$\boldsymbol{\sigma} = \nabla_\gamma \mathrm{W}\ \text{and}\ \boldsymbol{\mu} = \nabla_\chi \mathrm{W}. \tag{14}$$

The stress energy is given as

$$U_K = \int_{B_0} \Phi\{\boldsymbol{\sigma}, \boldsymbol{\mu}\} dv, \tag{15}$$

where

$$\begin{aligned} \Phi\{\boldsymbol{\sigma}, \boldsymbol{\mu}\} = &\frac{\mu' + \alpha'}{2}\sigma_{ij}\sigma_{ij} + \frac{\mu' - \alpha'}{2}\sigma_{ij}\sigma_{ji} + \frac{\lambda'}{2}\sigma_{kk}\sigma_{nn} \\ &+ \frac{\gamma' + \varepsilon'}{2}\mu_{ij}\mu_{ij} + \frac{\gamma' - \varepsilon'}{2}\mu_{ij}\mu_{ji} + \frac{\beta'}{2}\mu_{kk}\mu_{nn}, \end{aligned} \tag{16}$$

and the relations Eqs. (6)–(7) can be written as [12].

$$\boldsymbol{\gamma} = \frac{\partial \Phi}{\partial \boldsymbol{\sigma}},\ \text{and}\ \boldsymbol{\chi} = \frac{\partial \Phi}{\partial \boldsymbol{\mu}}. \tag{17}$$

We consider the work done by the stresses $\boldsymbol{\sigma}$ and $\boldsymbol{\mu}$ over the strains $\boldsymbol{\gamma}$ and $\boldsymbol{\chi}$ as in [13].

$$U = \int_{B_0} [\boldsymbol{\sigma} \cdot \boldsymbol{\gamma} + \boldsymbol{\mu} \cdot \boldsymbol{\chi}] dv \tag{18}$$

and

$$U = U_K = U_C \tag{19}$$

Here $\boldsymbol{\sigma} \cdot \boldsymbol{\gamma} = \sigma_{ji}\gamma_{ji}$ and $\boldsymbol{\mu} \cdot \boldsymbol{\chi} = \mu_{ji}\chi_{ji}$.
The stored kinetic energy T_C is defined as

$$T_C = \int_{B_0} \Upsilon_C dv = \frac{1}{2} \int_{B_0} \left(\rho \left(\frac{\partial \mathbf{u}}{\partial t} \right)^2 + \mathbf{J} \left(\frac{\partial \boldsymbol{\phi}}{\partial t} \right)^2 \right) dv, \tag{20}$$

The kinetic energy T_K is given as

$$T_K = \int_{B_0} \Upsilon_K \{\mathbf{p}, \mathbf{q}\} dv = \frac{1}{2} \int_{B_0} (\mathbf{p}^2 \rho^{-1} + \mathbf{q}^2 \mathbf{J}^{-1}) dv, \tag{21}$$

where

$$\mathbf{p} = \frac{\partial \Upsilon_C}{\partial \dot{\mathbf{u}}} = \rho \frac{\partial \mathbf{u}}{\partial t} \text{ and } \mathbf{q} = \frac{\partial \Upsilon_C}{\partial \dot{\boldsymbol{\phi}}} = \mathbf{J} \frac{\partial \boldsymbol{\phi}}{\partial t}, \tag{22}$$

and

$$\frac{\partial \mathbf{u}}{\partial t} = \frac{\partial \Upsilon_K}{\partial \mathbf{p}} = \mathbf{p} \rho^{-1} \text{ and } \frac{\partial \boldsymbol{\phi}}{\partial t} = \frac{\partial \Upsilon_K}{\partial \mathbf{q}} = \mathbf{q} \mathbf{J}^{-1}, \tag{23}$$

The work T_W done by the inertia forces over displacement and microrotation is given as in [12].

$$T_W = \int_{B_0} \Upsilon_W dv = \int_{B_0} \left(\frac{\partial \mathbf{p}}{\partial t} \cdot \mathbf{u} + \frac{\partial \mathbf{q}}{\partial t} \cdot \boldsymbol{\phi} \right) dv \tag{24}$$

Keeping in mind that the variation of $\mathbf{p}$ $\mathbf{u}$, $\mathbf{q}$, $\boldsymbol{\phi}$, $\delta \mathbf{u}$, and $\delta \boldsymbol{\phi}$ is zero at t_0 and t_k we can integrate by parts

$$\int_{t_0}^{t_k} T_K dt = \int_{t_0}^{t_k} T_I dt = \frac{1}{2} \int_{B_0} (\mathbf{p} \cdot \mathbf{u} + \mathbf{q} \cdot \boldsymbol{\phi}) dv \Big|_{t_0}^{t_k} - \int_{t_0}^{t_k} T_W dt \tag{25}$$

$$\delta \int_{t_0}^{t_k} T_K = -\delta \int_{t_0}^{t_k} T_W \tag{26}$$

or

$$\delta T_C = \delta T_K = -\delta T_W \tag{27}$$

and therefore

$$\int_{t_0}^{t_k} \int_{B_0} \left(\left(\mathbf{p} \cdot \delta \left(\frac{\partial \mathbf{u}}{\partial t} \right) + \mathbf{q} \cdot \delta \left(\frac{\partial \phi}{\partial t} \right) \right) \right) dv dt = - \int_{t_0}^{t_k} \int_{B_0} \left(\frac{\partial \mathbf{p}}{\partial t} \cdot \delta \mathbf{u} + \frac{\partial \mathbf{q}}{\partial t} \cdot \delta \boldsymbol{\phi} \right) dv dt \tag{28}$$

2.3 Variational principle for elastodynamics

We modify the HPR principle [14] for the case of Cosserat elastodynamics in the following way: for any set $\mathcal{A}$v of all admissible states $\mathfrak{s} = [\mathbf{u}, \boldsymbol{\phi}, \boldsymbol{\gamma}, \boldsymbol{\chi}, \boldsymbol{\sigma}, \boldsymbol{\mu}]$ that satisfy the strain-displacement and torsion-rotation relations Eq. (5), the zero variation

$$\delta\Theta(\mathfrak{s}) = 0$$

of the functional

$$\Theta(\mathfrak{s}) = \int_{t_0}^{t_k} \left[U_K + T_C - \int_{B_0} \left(\boldsymbol{\sigma} \cdot \boldsymbol{\gamma} + \boldsymbol{\mu} \cdot \boldsymbol{\chi} + \mathbf{p} \frac{\partial \mathbf{u}}{\partial t} + \mathbf{q} \frac{\partial \boldsymbol{\phi}}{\partial t} \right) dv \right] dt$$
$$+ \int_{t_0}^{t} \int_{\mathcal{G}_1} [\boldsymbol{\sigma}_\mathbf{n} \cdot (\mathbf{u} - \mathbf{u}_0) + \boldsymbol{\mu}_\mathbf{n} (\boldsymbol{\phi} - \boldsymbol{\phi}_0)] da dt + \int_{t_0}^{t} \int_{\mathcal{G}_2} [\boldsymbol{\sigma}_0 \cdot \mathbf{u} + \boldsymbol{\mu}_0 \cdot \boldsymbol{\phi}] da dt \tag{29}$$

at $\mathfrak{s} \in \mathcal{A}$ is equivalent of $\mathfrak{s}$ to be a solution of the system of equilibrium Eqs. (1)–(2), constitutive relations Eqs. (6)–(7), which satisfies the mixed boundary conditions Eqs. (8)–(9).

Proof of the variational principle for elastodynamics

Let us consider the variation of the functional $\Theta(\mathfrak{s})$:

$$\delta\Theta(\mathfrak{s}) = \int_{t_0}^{t_k} [\delta U_K + \delta T_C] dt$$
$$- \int_{t_0}^{t_k} \int_{B_0} \left[\delta\boldsymbol{\sigma} \cdot \boldsymbol{\gamma} + \boldsymbol{\sigma} \cdot \delta\boldsymbol{\gamma} + \delta\boldsymbol{\mu} \cdot \boldsymbol{\chi} + \boldsymbol{\mu} \cdot \delta\boldsymbol{\chi} + \frac{\partial \mathbf{u}}{\partial t} \delta\mathbf{p} + \mathbf{p} \cdot \delta\left(\frac{\partial \mathbf{u}}{\partial t}\right) + \delta\mathbf{q} \cdot \frac{\partial \boldsymbol{\phi}}{\partial t} + \mathbf{q} \cdot \delta\left(\frac{\partial \boldsymbol{\phi}}{\partial t}\right) \right] dv dt$$
$$+ \int_{t_0}^{t} \int_{\mathcal{G}_1} [\delta\boldsymbol{\sigma}_n \cdot (\mathbf{u} - \mathbf{u}_0) + \boldsymbol{\sigma}_n \delta\mathbf{u} + \delta\boldsymbol{\mu}_n \cdot (\boldsymbol{\phi} - \boldsymbol{\phi}_0) + \boldsymbol{\mu}_n \delta\boldsymbol{\phi}] da dt$$
$$+ \int_{t_0}^{t} \int_{\mathcal{G}_2} [\boldsymbol{\sigma}_0 \cdot \delta\mathbf{u} + \boldsymbol{\mu}_0 \cdot \delta\boldsymbol{\phi}] da dt$$

Taking into account Eq. (5) we can perform the integration by parts

$$\int_{B_0} \boldsymbol{\sigma} \cdot \delta\boldsymbol{\gamma} dv = \int_{\partial B_0} \boldsymbol{\sigma}_n \cdot \delta\mathbf{u} da - \int_{B_0} \delta\mathbf{u} \cdot \mathbf{div}\boldsymbol{\sigma} dv + \int_{B_0} \boldsymbol{\varepsilon}\boldsymbol{\sigma} \cdot \delta\boldsymbol{\phi} dv$$

$$\int_{B_0} \boldsymbol{\mu} \cdot \delta\boldsymbol{\chi} dv = \int_{\partial B_0} \boldsymbol{\mu}_n \cdot \delta\boldsymbol{\phi} da - \int_{B_0} \delta\boldsymbol{\phi} \cdot \mathbf{div}\boldsymbol{\mu} dv$$

and based on Eqs. (17)–(23)

$$\delta\Phi = \frac{\partial\Phi}{\partial\boldsymbol{\sigma}} \cdot \delta\boldsymbol{\sigma} + \frac{\partial\Phi}{\partial\boldsymbol{\mu}} \cdot \delta\boldsymbol{\mu}, \quad \delta\Upsilon_C = \frac{\partial\Upsilon_C}{\partial\left(\frac{\partial\mathbf{u}}{\partial t}\right)} \cdot \delta\left(\frac{\partial \mathbf{u}}{\partial t}\right) + \frac{\partial\Upsilon_C}{\partial\left(\frac{\partial\boldsymbol{\phi}}{\partial t}\right)} \cdot \delta\left(\frac{\partial \boldsymbol{\phi}}{\partial t}\right).$$

Then keeping in mind that $\delta T_K = -\delta T$ and Eq. (28) we can rewrite the expression for the variation of the functional $\delta\Theta(\mathfrak{s})$ in the following form

$$
\begin{aligned}
\delta\Theta(\mathfrak{s}) = & \int_{t_0}^{t}\int_{B_0}\left[\left(\frac{\partial\Phi}{\partial\boldsymbol{\sigma}}-\boldsymbol{\gamma}\right)\cdot\delta\boldsymbol{\sigma}\right]dvdt+\int_{t_0}^{t}\int_{B_0}\left[\left(\frac{\partial\Phi}{\partial\boldsymbol{\mu}}-\boldsymbol{\chi}\right)\cdot\delta\boldsymbol{\mu}\right]dvdt \\
& +\int_{t_0}^{t}\int_{B_0}\left[\left(\rho\frac{\partial\mathbf{u}}{\partial t}-\mathbf{p}\right)\cdot\delta\left(\frac{\partial\mathbf{u}}{\partial t}\right)\right]dvdt+\int_{t_0}^{t}\int_{B_0}\left[\left(\mathbf{J}\frac{\partial\varphi}{\partial t}-\mathbf{q}\right)\cdot\delta\left(\frac{\partial\varphi}{\partial t}\right)\right]dvdt \\
& +\int_{t_0}^{t}\int_{B_0}\left[\left(\mathbf{div}\boldsymbol{\sigma}-\frac{\partial\mathbf{p}}{\partial t}\right)\cdot\delta\mathbf{u}\right]dvdt+\int_{t_0}^{t}\int_{B_0}\left[\left(\mathbf{div}\boldsymbol{\mu}+\boldsymbol{\varepsilon}\cdot\boldsymbol{\sigma}-\frac{\partial\mathbf{q}}{\partial t}\right)\cdot\delta\mathbf{u}\right]dvdt \\
& +\int_{t_0}^{t}\int_{\mathcal{G}_1}[(\mathbf{u}-\mathbf{u_0})\cdot\delta\boldsymbol{\sigma}_{\mathbf{n}}]dadt+\int_{t_0}^{t}\int_{\mathcal{G}_1}[(\boldsymbol{\phi}-\boldsymbol{\phi}_{\mathbf{0}})\cdot\delta\boldsymbol{\mu}_{\mathbf{n}}]dadt \\
& +\int_{t_0}^{t}\int_{\mathcal{G}_2}[(\boldsymbol{\sigma}_{\mathbf{n}}-\boldsymbol{\sigma}_{\mathbf{0}})\cdot\delta\mathbf{u}]dadt+\int_{t_0}^{t}\int_{\mathcal{G}_2}[(\boldsymbol{\mu}_{\mathbf{n}}-\boldsymbol{\mu}_{\mathbf{0}})\cdot\delta\boldsymbol{\phi}]dadt
\end{aligned}
$$

3. Dynamic Cosserat plate theory

In this section we review our stress, couple stress and kinematic assumptions of the Cosserat plate [7]. We consider the thin plate P, where h is the thickness of the plate and $x_3 = 0$ represents its middle plane. The sets T and B are the top and bottom surfaces contained in the planes $x_3 = h/2$, $x_3 = -h/2$ respectively and the curve Γ is the boundary of the middle plane of the plate.

The set of points $P = \left(\Gamma \times \left[-\frac{h}{2},\frac{h}{2}\right]\right) \cup T \cup B$ forms the entire surface of the plate and $\Gamma_u \times \left[-\frac{h}{2}, \frac{h}{2}\right]$ is the lateral part of the boundary where displacements and microrotations are prescribed. The notation $\Gamma_\sigma = \Gamma\backslash\Gamma_u$ of the remainder we use to describe the lateral part of the boundary edge $\Gamma_\sigma \times \left[-\frac{h}{2}, \frac{h}{2}\right]$ where stress and couple stress are prescribed. We also use notation P_0 for the middle plane internal domain of the plate.

In our case we consider the vertical load and pure twisting momentum boundary conditions at the top and bottom of the plate, which can be written in the form:

$$\sigma_{33}(x_1,x_2,h/2,t) = \sigma^t(x_1,x_2,t), \sigma_{33}(x_1,x_2,-h/2,t) = \sigma^b(x_1,x_2,t), \tag{30}$$

$$\sigma_{3\beta}(x_1,x_2,\pm h/2,t) = 0, \tag{31}$$

$$\mu_{33}(x_1,x_2,h/2,t) = \mu^t(x_1,x_2,t), \mu_{33}(x_1,x_2,-h/2,t) = \mu^b(x_1,x_2,t), \tag{32}$$

$$\mu_{3\beta}(x_1,x_2,\pm h/2,t) = 0, \tag{33}$$

where $(x_1,x_2) \in P_0$.

We will also consider the rotatory inertia $\mathbf{J}$ in the form

$$\mathbf{J} = \begin{pmatrix} J_{11} & J_{12} & 0 \\ J_{12} & J_{22} & 0 \\ 0 & 0 & J_{33} \end{pmatrix}$$

Let $\mathcal{A}$ denote the set of all admissible states that satisfy the Cosserat plate strain-displacement relation Eq. (5) and let Θ be a functional on $\mathcal{A}$ defined by

$$\Theta(s,\eta) = U_K^S + T_C^S - \int\limits_{P_0}\left(\mathcal{S}\cdot\mathcal{E} + \mathbf{P}\cdot\frac{\partial\mathcal{U}}{\partial t} - \hat{P}\cdot\mathcal{W} + \upsilon\Omega_3^0\right)da + \int_{\Gamma_\sigma}\mathcal{S}_o\cdot(\mathcal{U}-\mathcal{U}_o)ds + \int_{\Gamma_u}\mathcal{S}_n\cdot\mathcal{U}ds, \tag{34}$$

for every $s = [\mathcal{U}, \mathcal{E}, \mathcal{S}] \in \mathcal{A}$. Here $\hat{P} = (\hat{p}_1, \hat{p}_2)$ and $\mathcal{W} = (W, W^*)$, $\hat{p}_1 = \eta p$ and $\hat{p}_2 = \frac{2}{3}(1-\eta)p$

Here the plate stress and kinetic energy density by the formulas

$$U_K^{\mathcal{S}} = \int_{P_0} \Phi(\mathcal{S}) da, \; T_K^{\mathcal{S}} = \int_{P_0} \Upsilon_C\left(\frac{\partial \mathcal{U}}{\partial t}\right) da \tag{35}$$

where P_0 is the internal domain of the middle plane of the plate.

$$\begin{aligned}
\Phi(\mathcal{S}) = &-\frac{3\lambda(M_{\alpha\alpha})(M_{\beta\beta})}{h^3\mu(3\lambda+2\mu)} \\
&+\frac{3(\alpha+\mu)M_{\alpha\beta}^2}{2h^3\alpha\mu} + \frac{3(\alpha+\mu)}{160h^3\alpha\mu}\left[8\hat{Q}_\alpha\hat{Q}_\alpha + 15Q_\alpha\hat{Q}_\alpha + 20\hat{Q}_\alpha Q_\alpha^* + 8Q_\alpha^* Q_\alpha^*\right] \\
&+\frac{3(\alpha-\mu)M_{\alpha\beta}^2}{2h^3\alpha\mu} + \frac{\alpha-\mu}{280h^3\alpha\mu}\left[21Q_\alpha\left(5\hat{Q}_\alpha + 4Q_\alpha^*\right)\right] \\
&-\frac{\gamma-\varepsilon}{160h\gamma\varepsilon}\left[24R_{\alpha\alpha}^2 + 45R_{\alpha\alpha}^* + 60R_{\alpha\beta}R_{\alpha\beta}^* + 48R_{12}R_{21}\right] \\
&+\frac{3(\gamma+\varepsilon)S_\alpha^* S_\alpha^*}{2h^3\gamma\varepsilon} + \frac{\gamma+\varepsilon}{160h^3\gamma\varepsilon}\left[8R_{\alpha\beta}^2 + 15R_{\alpha\beta}^* R_{\alpha\beta}^* + 20R_{\alpha\beta}R_{\alpha\beta}^*\right] \\
&-\frac{3\beta}{80h\gamma(3\beta+2\gamma)}\left[-8(R_{\alpha\alpha})(R_{\beta\beta}) - 15(R_{\alpha\alpha}^*)\left(R_{\beta\beta}^*\right) - 20(R_{\alpha\alpha})(R_{\alpha\alpha}^*)\right] \\
&-\frac{\beta}{4\gamma(3\beta+2\gamma)}\left[(2R_{\alpha\alpha} + 3R_{\alpha\alpha}^*)t - h\left(V^2+T^2\right)\right] \\
&+\frac{\lambda}{560h\mu(3\lambda+2\mu)}\left[\frac{5+3\eta}{(1+\eta)}pM_{\alpha\alpha}\right] \\
&+\frac{(\lambda+\mu)h}{840\mu(3\lambda+2\mu)}\left(\frac{140+168\eta+51\eta^2}{4(1+\eta)^2}\right)p^2 + \frac{(\lambda+\mu)h}{2\mu(3\lambda+2\mu)}\sigma_0^2 + \frac{\varepsilon h}{12h\gamma\varepsilon}\left[(3T^2+V^2)\right]
\end{aligned} \tag{36}$$

and

$$\begin{aligned}
\Upsilon_C\left(\frac{\partial \mathcal{U}}{\partial t}\right) &- \frac{h\rho}{2}\left(\frac{\partial W}{\partial t}\right)^2 + \frac{4h\rho}{15}\left(\frac{\partial W^*}{\partial t}\right)^2 + \frac{2h\rho}{3}\left(\frac{\partial W}{\partial t}\frac{\partial W^*}{\partial t}\right) + \frac{h^3\rho}{24}\left(\frac{\partial \Psi_\alpha}{\partial t}\right)^2 \\
&+\frac{4hJ_{\alpha\beta}}{15}\left(\frac{\partial \Omega_\alpha^0}{\partial t}\frac{\partial \Omega_\beta^0}{\partial t}\right) + \frac{hJ_{\alpha\beta}}{2}\left(\frac{\partial \hat{\Omega}_\alpha^0}{\partial t}\frac{\partial \hat{\Omega}_\beta^0}{\partial t}\right) + \frac{2hJ_{\alpha\beta}}{3}\left(\frac{\partial \Omega_\alpha^0}{\partial t}\frac{\partial \hat{\Omega}_\beta^0}{\partial t}\right) + \frac{hJ_{33}}{6}\left(\frac{\partial \Omega_3}{\partial t}\right)^2.
\end{aligned}$$

$\mathcal{S}$, $\mathcal{U}$ and $\mathcal{E}$ are the Cosserat plate stress, displacement and strain sets

$$\mathcal{S} = \left[M_{\alpha\beta}, Q_\alpha, Q_\alpha^*, \hat{Q}_\alpha, R_{\alpha\beta}, R_{\alpha\beta}^*, S_\beta^*\right], \tag{37}$$

$$\mathcal{S}_n = \left[\breve{M}_\alpha, \breve{Q}^*, \hat{\breve{Q}}_\alpha, \breve{R}_\alpha, \breve{R}_\alpha^*, \breve{S}^*\right], \tag{38}$$

$$\mathcal{S}_o = \left[\Pi_{o\alpha}, \Pi_{o3}, \Pi_{o3}^*, M_{o\alpha}, M_{o\alpha}^*, M_{o3}^*\right], \tag{39}$$

$$\mathcal{U} = \left[\Psi_\alpha, W, \Omega_3, \Omega_\alpha^0, W^*, \Omega_\alpha^0\right], \tag{40}$$

$$\mathcal{E} = \left[e_{\alpha\beta}, \omega_\beta, \omega_\alpha^*, \hat{\omega}_\alpha, \tau_{3\alpha}, \tau_{\alpha\beta}, \tau_{\alpha\beta}^*\right], \tag{41}$$

where $M_{\alpha\beta}n_\beta = \Pi_{o\alpha}$, $R_{\alpha\beta}n_\beta = M_{o\alpha}$, $Q_\alpha^* n_\alpha = \Pi_{o3}$, $S_\alpha^* n_\alpha = M_{o3}^*$, $\hat{Q}_\alpha n_\alpha = \Pi_{o3}^*$, $R_{\alpha\beta}^* n_\beta = M_{o\alpha}^*$, $\check{M}_\alpha = M_{\alpha\beta}n_\beta$, $\check{Q}^* = Q_\beta^* n_\beta$, $\check{R}_\alpha = R_{\alpha\beta}n_\beta$, $\check{S}^* = S_\beta^* n_\beta$, $\check{Q}^{\hat{}} = \check{Q}^{\hat{}}_\beta n_\beta$, $\check{R}_\alpha^* = \check{R}_{\alpha\beta}^* n_\beta$. ($n_\beta$ is the outward unit normal vector to Γ_u).

The plate characteristics provide the approximation of the components of the three-dimensional tensors σ_{ji} and μ_{ji}

$$\sigma_{\alpha\beta} = \frac{6}{h^2}\zeta M_{\alpha\beta}(x_1, x_2, t), \tag{42}$$

$$\sigma_{3\beta} = \frac{3}{2h}(1-\zeta^2)Q_\beta(x_1, x_2, t), \tag{43}$$

$$\sigma_{\beta 3} = \frac{3}{2h}(1-\zeta^2)Q_\beta^*(x_1, x_2, t) + \frac{3}{2h}\hat{Q}_\beta(x_1, x_2, t), \tag{44}$$

$$\sigma_{33} = -\frac{3}{4}\left(\frac{1}{3}\zeta^3 - \zeta\right)p_1(x_1, x_2, t) + \zeta p_2(x_1, x_2, t) + \sigma_0(x_1, x_2, t), \tag{45}$$

$$\mu_{\alpha\beta} = \frac{3}{2h}(1-\zeta^2)R_{\alpha\beta}(x_1, x_2, t) + \frac{3}{2h}R_{\alpha\beta}^*(x_1, x_2, t), \tag{46}$$

$$\mu_{\beta 3} = \frac{6}{h^2}\zeta S_\beta^*(x_1, x_2, t), \tag{47}$$

$$\mu_{3\beta} = 0, \tag{48}$$

$$\mu_{33} = \zeta V(x_1, x_2, t) + T(x_1, x_2, t), \tag{49}$$

where

$$p(x_1, x_2, t) = \sigma^t(x_1, x_2, t) - \sigma^b(x_1, x_2, t), \tag{50}$$

$$\sigma_0(x_1, x_2, t) = \frac{1}{2}\left(\sigma^t(x_1, x_2, t) + \sigma^b(x_1, x_2, t)\right), \tag{51}$$

$$V(x_1, x_2, t) = \frac{1}{2}\left(\mu^t(x_1, x_2, t) - \mu^b(x_1, x_2, t)\right), \tag{52}$$

$$T(x_1, x_2, t) = \frac{1}{2}\left(\mu^t(x_1, x_2, t) + \mu^b(x_1, x_2, t)\right). \tag{53}$$

The pressures p_1 and p_2 are chosen in the form

$$p_1(x_1, x_2, t) = \eta p(x_1, x_2, t), \tag{54}$$

$$p_2(x_1, x_2, t) = \frac{(1-\eta)}{2}p(x_1, x_2, t). \tag{55}$$

and $\eta \in \mathbb{R}$ is called *splitting parameter*.
The three-dimensional displacements u_i and microrotations ϕ_i

$$u_\alpha = \frac{h}{2}\zeta\Psi_\alpha(x_1, x_2, t), \tag{56}$$

$$u_3 = W(x_1, x_2, t) + (1-\zeta^2)W^*(x_1, x_2, t), \tag{57}$$

$$\phi_\alpha = \Omega_\alpha^0(x_1, x_2, t)(1-\zeta^2) + \hat{\Omega}_\alpha(x_1, x_2, t), \tag{58}$$

$$\phi_3 = \zeta\Omega_3(x_1, x_2, t), \tag{59}$$

and the three-dimensional strain and torsion tensors γ_{ji} and χ_{ji}

$$\gamma_{\alpha\beta} = \frac{6}{h^2}\zeta e_{\alpha\beta}(x_1, x_2, t), \tag{60}$$

$$\gamma_{3\beta} = \frac{3}{2h}\left(1-\zeta^2\right)\omega_\beta(x_1, x_2, t), \tag{61}$$

$$\gamma_{\beta 3} = \frac{3}{2h}\left(1-\zeta^2\right)\omega^*_\beta(x_1, x_2, t) + \frac{3}{2h}\hat{\omega}_\beta(x_1, x_2, t), \tag{62}$$

$$\chi_{\alpha\beta} = \frac{3}{2h}\left(1-\zeta^2\right)\tau_{\alpha\beta}(x_1, x_2, t) + \frac{3}{2h}\tau^*_{\alpha\beta}(x_1, x_2, t), \tag{63}$$

$$\chi_{3\beta} = \frac{6}{h^2}\zeta\tau^*_\beta(x_1, x_2, t), \tag{64}$$

where $\zeta = \frac{2x_3}{h}$.

Then zero variation of the functional

$$\delta\Theta(s, \eta) = 0$$

is equivalent to the plate bending system of equations (A) and constitutive formulas (B) mixed problems.

A. The bending equilibrium system of equations:

$$M_{\alpha\beta,\,\alpha} - Q_\beta = I_1\frac{\partial^2\Psi_\beta}{\partial t^2}, \tag{65}$$

$$Q^*_{\alpha,\,\alpha} + \hat{p}_1 = I_2\frac{\partial^2 W^*}{\partial t^2}, \tag{66}$$

$$R_{\alpha\beta,\,\alpha} + \varepsilon_{3\beta\gamma}\left(Q^*_\gamma - Q_\gamma\right) = I_{\alpha\beta}\frac{\partial^2\Omega^0_\alpha}{\partial t^2}, \tag{67}$$

$$\varepsilon_{3\beta\gamma}M_{\beta\gamma} + S^*_{\alpha,\,\alpha} = I_3\frac{\partial^2\Omega_3}{\partial t^2}, \tag{68}$$

$$\hat{Q}_{\alpha,\,\alpha} + \hat{p}_2 = I_2\frac{\partial^2 W}{\partial t^2}, \tag{69}$$

$$R^*_{\alpha\beta,\,\alpha} + \varepsilon_{3\beta\gamma}\hat{Q}_\gamma = I^0_{\alpha\beta}\frac{\partial^2\hat{\Omega}^0_\alpha}{\partial t^2}, \tag{70}$$

where $I_1 = \frac{h^3}{12}\rho$, $I_2 = \frac{2h}{3}\rho$, $I_{\alpha\beta} = \frac{5h}{6}J_{\alpha\beta}$, $I_3 = \frac{h^2}{6}J_{33}$, $I^0_{\alpha\beta} = \frac{2h}{3}J_{\alpha\beta}$, $\hat{p}_1 = \eta_{opt}p$, and $\hat{p}_2 = \frac{2}{3}\left(1-\eta_{opt}\right)p$, with the resultant traction boundary conditions:

$$M_{\alpha\beta}n_\beta = \Pi_{o\alpha},\quad R_{\alpha\beta}n_\beta = M_{o\alpha}, \tag{71}$$

$$Q^*_\alpha n_\alpha = \Pi_{o3},\quad S^*_\alpha n_\alpha = \Upsilon_{o3}, \tag{72}$$

at the part Γ_σ and the resultant displacement boundary conditions

$$\Psi_\alpha = \Psi_{o\alpha}, W = W_o, \Omega^0_\alpha = \Omega^0_{o\alpha}, \Omega_3 = \Omega_{o3}, \tag{73}$$

at the part Γ_u.

B. Constitutive formulas in the reverse form:[1]

$$M_{\alpha\alpha} = \frac{\mu(\lambda+\mu)h^3}{3(\lambda+2\mu)}\Psi_{\alpha,\alpha} + \frac{\lambda\mu h^3}{6(\lambda+2\mu)}\Psi_{\beta,\beta} + \frac{(3p_1+5p_2)\lambda h^2}{30(\lambda+2\mu)}, \tag{74}$$

$$M_{\beta\alpha} = \frac{(\mu-\alpha)h^3}{12}\Psi_{\alpha,\beta} + \frac{(\mu+\alpha)h^3}{12}\Psi_{\beta,\alpha} + (-1)^{\alpha'}\frac{\alpha h^3}{6}\Omega_3, \tag{75}$$

$$R_{\beta\alpha} = \frac{5(\gamma-\varepsilon)h}{6}\Omega^0_{\beta,\alpha} + \frac{5(\gamma+\varepsilon)h}{6}\Omega^0_{\alpha,\beta}, \tag{76}$$

$$R_{\alpha\alpha} = \frac{10h\gamma(\beta+\gamma)}{3(\beta+2\gamma)}\Omega^0_{\alpha,\alpha} + \frac{5h\beta\gamma}{3(\beta+2\gamma)}\Omega^0_{\beta,\beta}, \tag{77}$$

$$R^*_{\beta\alpha} = \frac{2(\gamma-\varepsilon)h}{3}\hat{\Omega}_{\beta,\alpha} + \frac{2(\gamma+\varepsilon)h}{3}\hat{\Omega}_{\alpha,\beta}, \tag{78}$$

$$R^*_{\alpha\alpha} = \frac{8\gamma(\gamma+\beta)h}{3(\beta+2\gamma)}\hat{\Omega}_{\alpha,\alpha} + \frac{4\gamma\beta h}{3(\beta+2\gamma)}\hat{\Omega}_{\beta,\beta}, \tag{79}$$

$$Q_\alpha = \frac{5(\mu+\alpha)h}{6}\Psi_\alpha + \frac{5(\mu-\alpha)h}{6}W_{,\alpha} + \frac{2(\mu-\alpha)h}{3}W^*_{,\alpha} + (-1)^\beta\frac{5h\alpha}{3}\Omega^0_\beta + (-1)^\beta\frac{5h\alpha}{3}\hat{\Omega}_\beta, \tag{80}$$

$$Q^*_\alpha = \frac{5(\mu-\alpha)h}{6}\Psi_\alpha + \frac{5(\mu-\alpha)^2h}{6(\mu+\alpha)}W_{,\alpha} + \frac{2(\mu+\alpha)h}{3}W^*_{,\alpha} + (-1)^\alpha\frac{5h\alpha}{3}\left(\Omega^0_\beta + \frac{(\mu-\alpha)}{(\mu+\alpha)}\hat{\Omega}_\beta\right), \tag{81}$$

$$\hat{Q}_\alpha = \frac{8\alpha\mu h}{3(\mu+\alpha)}W_{,\alpha} + (-1)^\alpha\frac{8\alpha\mu h}{3(\mu+\alpha)}\hat{\Omega}_\beta, \tag{82}$$

$$S^*_\alpha = \frac{5\gamma\varepsilon h^3}{3(\gamma+\varepsilon)}\Omega_{3,\alpha}, \tag{83}$$

and the optimal value η_{opt} of the splitting parameter is given as in [10]

$$\eta_{opt} = \frac{2\mathcal{W}^{(00)} - \mathcal{W}^{(10)} - \mathcal{W}^{(01)}}{2\left(\mathcal{W}^{(11)} + \mathcal{W}^{(00)} - \mathcal{W}^{(10)} - \mathcal{W}^{(01)}\right)}. \tag{84}$$

where

$$\mathcal{W}^{(ij)} = \mathcal{S}|_{\eta=i} \cdot \mathcal{E}|_{\eta=j}.$$

We also assume that the initial condition can be presented in the form

$$\mathcal{U}(x_1,x_2,0) = \mathcal{U}^0(x_1,x_2),\ \frac{\partial\mathcal{U}}{\partial t}(x_1,x_2,0) = \mathcal{V}^0(x_1,x_2)$$

4. Cosserat plate dynamic field equations

The Cosserat plate field equations are obtained by substituting the relations Eqs. (74)–(83) into the system of Eqs. (65)–(70) similar to [10]:

[1] In the following formulas a subindex $\beta = 1$ if $\alpha = 2$ and $\beta = 2$ if $\alpha = 1$.

$$L\mathcal{U} = K\frac{\partial^2 \mathcal{U}}{\partial t^2} + \mathcal{F}(\eta), \tag{85}$$

where

$$L = \begin{bmatrix} L_{11} & L_{12} & L_{13} & L_{14} & 0 & L_{16} & kL_{13} & 0 & L_{16} \\ L_{12} & L_{22} & L_{23} & L_{24} & L_{16} & 0 & kL_{23} & L_{16} & 0 \\ -L_{13} & -L_{23} & L_{33} & 0 & L_{35} & L_{36} & L_{77} & L_{38} & L_{39} \\ L_{41} & L_{42} & 0 & L_{44} & 0 & 0 & 0 & 0 & 0 \\ 0 & -L_{16} & -L_{38} & 0 & L_{55} & L_{56} & -kL_{35} & L_{58} & 0 \\ L_{16} & 0 & -L_{39} & 0 & L_{56} & L_{66} & -kL_{36} & 0 & L_{58} \\ -L_{13} & -L_{14} & L_{73} & 0 & L_{35} & L_{36} & L_{77} & L_{78} & L_{79} \\ 0 & -L_{16} & -L_{78} & 0 & L_{85} & L_{56} & -kL_{35} & L_{88} & kL_{56} \\ L_{16} & 0 & -L_{79} & 0 & L_{56} & L_{55} & -kL_{36} & kL_{56} & L_{99} \end{bmatrix},$$

$$K - \begin{bmatrix} \frac{h^3}{12}\rho & 0 & 0 & 0 & 0 & 0 & 0 & 0 & 0 \\ 0 & \frac{h^3}{12}\rho & 0 & 0 & 0 & 0 & 0 & 0 & 0 \\ 0 & 0 & \frac{2h}{3}\rho & 0 & 0 & 0 & 0 & 0 & 0 \\ 0 & 0 & 0 & \frac{h^2}{6}J_{33} & 0 & 0 & 0 & 0 & 0 \\ 0 & 0 & 0 & 0 & \frac{5h}{6}J_{11} & \frac{5h}{6}J_{12} & 0 & 0 & 0 \\ 0 & 0 & 0 & 0 & \frac{5h}{6}J_{12} & \frac{5h}{6}J_{22} & 0 & 0 & 0 \\ 0 & 0 & 0 & 0 & 0 & 0 & \frac{2h}{3}\rho & 0 & 0 \\ 0 & 0 & 0 & 0 & 0 & 0 & 0 & \frac{2h}{3}J_{11} & \frac{2h}{3}J_{12} \\ 0 & 0 & 0 & 0 & 0 & 0 & 0 & \frac{2h}{3}J_{12} & \frac{2h}{3}J_{22} \end{bmatrix},$$

$$\mathcal{U} = \left[\Psi_1,\ \Psi_2,\ W,\ \Omega_3,\ \Omega_1^0,\ \Omega_2^0,\ W^*,\ \Omega_1^0,\ \Omega_2^0\right]^T,$$

$$\mathcal{F}(\eta) = \left[-\frac{3h^2\lambda\left(3p_{1,1}+5p_{2,1}\right)}{30(\lambda+2\mu)},\quad -\frac{3h^2\lambda\left(3p_{1,2}+5p_{2,2}\right)}{30(\lambda+2\mu)},\quad -p_1,\quad 0,\quad 0,\quad 0,\quad \frac{h^2\left(3p_1+4p_2\right)}{24},\quad 0,\quad 0\right]^T,$$

$$p_1 = \eta p, \quad p_2 = \frac{(1-\eta)}{2}p$$

The operators L_{ij} are given as follows

$$L_{11} = c_1\frac{\partial^2}{\partial x_1^2} + c_2\frac{\partial^2}{\partial x_2^2} - c_3, \quad L_{12} = (c_1 - c_2)\frac{\partial^2}{\partial x_1 x_2}, \quad L_{13} = c_{11}\frac{\partial}{\partial x_1},$$

$$L_{14} = c_{12}\frac{\partial}{\partial x_2}, \quad L_{16} = c_{13}, \quad L_{17} = k_1 c_{11}\frac{\partial}{\partial x_1},$$

$$L_{22} = c_2\frac{\partial^2}{\partial x_1^2} + c_1\frac{\partial^2}{\partial x_2^2} - c_3, \quad L_{23} = c_{11}\frac{\partial}{\partial x_2}, \quad L_{24} = -c_{12}\frac{\partial}{\partial x_1},$$

$$L_{33} = c_3\left(\frac{\partial^2}{\partial x_1^2} + \frac{\partial^2}{\partial x_2^2}\right), \quad L_{35} = -c_{13}\frac{\partial}{\partial x_2}, \quad L_{36} = c_{13}\frac{\partial}{\partial x_1},$$

$$L_{38} = -c_{10}\frac{\partial}{\partial x_2}, \quad L_{39} = c_{10}\frac{\partial}{\partial x_1}, \quad L_{41} = -c_{12}\frac{\partial}{\partial x_2},$$

$$L_{42} = c_{12}\frac{\partial}{\partial x_1}, \quad L_{44} = c_6\left(\frac{\partial^2}{\partial x_1^2} + \frac{\partial^2}{\partial x_2^2}\right) - 2c_{12}, \quad L_{55} = c_7\frac{\partial^2}{\partial x_1^2} + c_8\frac{\partial^2}{\partial x_2^2} - 2c_{13},$$

$$L_{56} = (c_7 - c_8)\frac{\partial^2}{\partial x_1 x_2}, \quad L_{58} = -c_9, \quad L_{66} = c_8\frac{\partial^2}{\partial x_1^2} + c_7\frac{\partial^2}{\partial x_2^2} - 2c_{13},$$

$$L_{73} = c_5\left(\frac{\partial^2}{\partial x_1^2} + \frac{\partial^2}{\partial x_2^2}\right), \quad L_{77} = c_4\left(\frac{\partial^2}{\partial x_1^2} + \frac{\partial^2}{\partial x_2^2}\right), \quad L_{78} = -c_{14}\frac{\partial}{\partial x_2},$$

$$L_{79} = c_{14}\frac{\partial}{\partial x_1}, \quad L_{85} = c_7\frac{\partial^2}{\partial x_1^2} + c_8\frac{\partial^2}{\partial x_2^2} - 2c_{13}, \quad L_{88} = c_7\frac{\partial^2}{\partial x_1^2} + c_8\frac{\partial^2}{\partial x_2^2} - c_{15},$$

$$L_{99} = c_8\frac{\partial^2}{\partial x_1^2} + c_7\frac{\partial^2}{\partial x_2^2} - c_{15}.$$

The coefficients c_i are given as

$$c_1 = \frac{h^3\mu(\lambda+\mu)}{3(\lambda+2\mu)}, \quad c_2 = \frac{h^3(\alpha+\mu)}{12}, \quad c_3 = \frac{5h(\alpha+\mu)}{6}, \quad c_4 = \frac{5h(\alpha-\mu)^2}{6(\alpha+\mu)},$$

$$c_5 = \frac{h(5\alpha^2+6\alpha\mu+5\mu^2)}{6(\alpha+\mu)}, \quad c_6 = \frac{h^3\gamma\varepsilon}{3(\gamma+\varepsilon)}, \quad c_7 = \frac{10h\gamma(\beta+\gamma)}{3(\beta+2\gamma)}, \quad c_8 = \frac{5h(\gamma+\varepsilon)}{6},$$

$$c_9 = \frac{10h\alpha^2}{3(\alpha+\mu)}, \quad c_{10} = \frac{5h\alpha(\alpha-\mu)}{3(\alpha+\mu)}, \quad c_{11} = \frac{5h(\alpha-\mu)}{6}, \quad c_{12} = \frac{h^3\alpha}{6},$$

$$c_{13} = \frac{5h\alpha}{3}, \quad c_{14} = \frac{h\alpha(5\alpha+3\mu)}{3(\alpha+\mu)}, \quad c_{15} = \frac{2h\alpha(5\alpha+4\mu)}{3(\alpha+\mu)}.$$

5. Numerical validation

For the validation purposes we provide the algorithm and computation results for the three-dimensional Cosserat elastodynamics. We also present the analysis of the numerical results based on the plate theory for the microelements of different shapes and orientations incorporated into the Cosserat plate.

5.1 Analysis of Cosserat plate vibrations based on the three-dimensional theory

In our computations we consider the plates made of polyurethane foam—a material reported in the literature to behave Cosserat like—and the values of the technical elastic parameters presented in [15]: $E = 299.5\,\text{MPa}$, $\nu = 0.44$, $l_t = 0.62\,\text{mm}$, $l_b = 0.327\,\text{mm}$, $N^2 = 0.04$. Taking into account that the ratio β/γ is equal to 1 for bending [15], these values of the technical constants correspond to the following values of Lamé and Cosserat parameters: $\lambda = 762.616\,\text{MPa}$, $\mu = 103.993\,\text{MPa}$, $\alpha = 4.333\,\text{MPa}$, $\beta = 39.975\,\text{MPa}$, $\gamma = 39.975\,\text{MPa}$, $\varepsilon = 4.505\,\text{MPa}$. We consider a low-density rigid foam usually characterized by the densities of 24–50 kg/m^3 [16]. In all further numerical computations we used the density value $\rho = 34$ kg/m^3 and different values the rotatory inertia $\mathbf{J}$.

Let us consider the plate B_0 being a rectangular cuboid $[0, a] \times 0, a] \times \left[-\frac{h}{2}, \frac{h}{2}\right]$. Let the sets T and B be the top and the bottom surfaces contained in the planes $x_3 = \frac{h}{2}$ and $x_3 = -\frac{h}{2}$ respectively, and the curve $\Gamma = \Gamma_1 \cup \Gamma_2$ be the lateral part of the boundary:

$$\Gamma_1 = \left\{ (x_1, x_2, x_3) : x_1 \in \{0, a\}, x_2 \in [0, a], x_3 \in \left[-\frac{h}{2}, \frac{h}{2}\right] \right\},$$

$$\Gamma_2 = \left\{ (x_1, x_2, x_3) : x_1 \in [0, a], x_2 \in \{0, a\}, x_3 \in \left[-\frac{h}{2}, \frac{h}{2}\right] \right\},$$

We solve the three-dimensional Cosserat equilibrium Eqs. (1)–(2) accompanied by the constitutive Eqs. (3)–(4) and strain-displacement and torsion-rotation relations Eq. (5) complemented by the following boundary conditions:

$$\Gamma_1 : u_2 = 0, u_3 = 0, \varphi_1 = 0, \sigma_{11} = 0, \mu_{12} = 0, \mu_{13} = 0; \tag{86}$$

$$\Gamma_2 : u_1 = 0, u_3 = 0, \varphi_2 = 0, \sigma_{22} = 0, \mu_{21} = 0, \mu_{23} = 0; \tag{87}$$

$$T : \sigma_{33} = p(x_1, x_2), \mu_{33} = 0; \tag{88}$$

$$B : \sigma_{33} = 0, \mu_{33} = 0. \tag{89}$$

where the initial distribution of the pressure is given as $p = \sin\left(\frac{\pi x_1}{a}\right) \sin\left(\frac{\pi x_2}{a}\right) \sin \omega t$ and the rotatory inertia tensor $\mathbf{J}$ is assumed to have a diagonal form

$$\mathbf{J} = \begin{bmatrix} J_x & 0 & 0 \\ 0 & J_y & 0 \\ 0 & 0 & J_z \end{bmatrix}. \tag{90}$$

Using the method of separation of variables and taking into account the boundary conditions Eqs. (86)–(87), we express the kinematic variables in the form:

$$u_1 = \cos\left(\frac{\pi x_1}{a}\right) \sin\left(\frac{\pi x_2}{a}\right) z_1(x_3) \sin \omega t, \tag{91}$$

$$u_2 = \sin\left(\frac{\pi x_1}{a}\right) \cos\left(\frac{\pi x_2}{a}\right) z_2(x_3) \sin \omega t, \tag{92}$$

$$u_3 = \sin\left(\frac{\pi x_1}{a}\right) \sin\left(\frac{\pi x_2}{a}\right) z_3(x_3) \sin \omega t, \tag{93}$$

$$\phi_1 = \sin\left(\frac{\pi x_1}{a}\right) \cos\left(\frac{\pi x_2}{a}\right) z_4(x_3) \sin \omega t, \tag{94}$$

$$\phi_2 = \cos\left(\frac{\pi x_1}{a}\right) \sin\left(\frac{\pi x_2}{a}\right) z_5(x_3) \sin \omega t, \tag{95}$$

$$\phi_3 = \cos\left(\frac{\pi x_1}{a}\right) \cos\left(\frac{\pi x_2}{a}\right) z_6(x_3) \sin \omega t, \tag{96}$$

where the functions $z_i(x_3)$ represent the transverse variations of the kinematic variables.

If we substitute the expressions Eqs. (91)–(96) into Eqs. (3)–(4) and then into Eqs. (1)–(2), we will obtain the following eigenvalue problem

$$\mathbf{B}z = \omega^2 \mathbf{A}z \tag{97}$$

where

$$\mathbf{B} = \begin{bmatrix} b_1L_2 + b_2L_0 & b_3L_0 & b_4L_1 & 0 & -b_5L_1 & -b_6L_0 \\ b_3L_0 & b_1L_2 + b_2L_0 & b_4L_1 & b_5L_1 & 0 & b_6L_0 \\ -b_4L_1 & b_4L_1 & b_7L_2 & b_6L_0 & -b_6L_0 & 0 \\ 0 & -b_5L_1 & b_6L_0 & b_9L_2 + b_{10}L_0 & b_{11}L_0 & b_{12}L_1 \\ b_5L_1 & 0 & -b_6L_0 & b_{11}L_0 & b_9L_2 + b_{10}L_0 & b_{12}L_1 \\ -b_6L_0 & b_6L_0 & 0 & -b_{12}L_1 & -b_{12}L_1 & b_{13}L_2 + b_2L_{14} \end{bmatrix}, \tag{98}$$

$$\mathbf{A} = \begin{bmatrix} -a^2\rho & 0 & 0 & 0 & 0 & 0 \\ 0 & -a^2\rho & 0 & 0 & 0 & 0 \\ 0 & 0 & -a^2\rho & 0 & 0 & 0 \\ 0 & 0 & 0 & -a^2J_x & 0 & 0 \\ 0 & 0 & 0 & 0 & -a^2J_y & 0 \\ 0 & 0 & 0 & 0 & 0 & -a^2J_z \end{bmatrix}, \tag{99}$$

$$z = \begin{bmatrix} z_1, & z_2, & z_3, & z_4, & z_5, & z_6 \end{bmatrix}^T, \tag{100}$$

and the differential operators L_i are defined as

$$L_0 = I, \quad L_1 = \frac{d}{dx_3}, \quad L_2 = \frac{d^2}{dx_3^2}$$

and the coefficients b_i are defined as

$$\begin{aligned} &b_1 = a^2(\mu + \alpha), && b_2 = -\pi^2(\alpha + \lambda + 3\mu), && b_3 = -\pi^2(\lambda + \mu - \alpha), \\ &b_4 = a\pi(\lambda + \mu - \alpha), && b_5 = 2a^2\alpha, && b_6 = 2a\pi\alpha, \\ &b_7 = a^2(2\mu + \lambda), && b_8 = -2\pi^2(\alpha + \mu), && b_9 = a^2(\gamma + \varepsilon), \\ &b_{10} = -\pi^2(\beta + \varepsilon + 3\gamma), && b_{11} = -\pi^2(\beta + \gamma - \varepsilon), && b_{12} = -a\pi(\beta + \gamma - \varepsilon), \\ &b_{13} = a^2(\beta + 2\gamma), && b_{14} = -2\pi^2(\gamma + \varepsilon) - 4a^2\alpha \end{aligned}$$

The system of differential Eq. (97) is complemented by the following boundary conditions $\mathbf{D}z = D_0$ for $x_3 = \frac{h}{2}$ and $\mathbf{D}z = 0$ for $x_3 = -\frac{h}{2}$.

$$\mathbf{D} = \begin{bmatrix} d_1L_1 & 0 & d_2L_0 & 0 & -d_3L_0 & 0 \\ 0 & d_1L_1 & d_2L_0 & d_3L_0 & 0 & 0 \\ d_4L_0 & d_4L_0 & d_5L_1 & 0 & 0 & 0 \\ 0 & 0 & 0 & d_6L_1 & 0 & d_7L_0 \\ 0 & 0 & 0 & 0 & d_6L_1 & d_7L_0 \\ 0 & 0 & 0 & d_8L_0 & d_8L_0 & d_9L_1 \end{bmatrix}, \tag{101}$$

$$D_0 = \begin{bmatrix} 0, & 0, & a, & 0, & 0, & 0 \end{bmatrix}^T, \tag{102}$$

and the coefficients d_i are defined as

$$\begin{aligned} &d_1 = a(\mu + \alpha), && d_2 = -\pi(\mu - \alpha), && d_3 = 2a\alpha, \\ &d_4 = a(\lambda + 2\mu), && d_5 = -\pi\lambda, && d_6 = a(\gamma + \varepsilon), \\ &d_7 = a(\gamma - \varepsilon), && d_8 = \pi\beta, && d_9 = a(\beta + 2\gamma). \end{aligned}$$

The idea for the solution of the eigenvalue problem Eq. (97) is based on the following algorithm:

Step 1. *Fix certain frequency value.*

We fix certain value of the frequency ω and force the Cosserat body to vibrate at this frequency.

Step 2. *Solve the three-dimensional Cosserat system of equations.*

Mathematically, fixing certain value of ω implies that three-dimensional system of Eq. (97) has a constant right-hand side and therefore can be solved for the kinematic variables as a static system of equations. We solve the system Eq. (97) using the high-precision Runge-Kutta method incorporated in Mathematica software similar to how it was done in [7].

Step 3. *Find large amplitudes of the kinematic variables.*

We run ω through an interval of positive real values and take note where the solution changes its sign and the amplitude of the solutions starts to grow indefinitely. This corresponds to the oscillation of the Cosserat body at its resonant frequency. Thus, when the frequency ω coincides with the natural frequency of the plate the resonance will occur and the large amplitude linear vibrations can be observed (**Figure 1**).

The comparison of the eigenfrequencies of the Cosserat plate with the eigenfrequencies of the three-dimensional Cosserat elasticity is given in the **Table 1**. The rotatory inertia principle moments used are $J_x = 0.001, J_y = 0.001, J_z = 0.001$, which represent a ball-shaped microelement (**Figure 2**). The relative error of the natural macro frequencies associated with the rotation of the middle plane and the flexural motion is less than 1%.

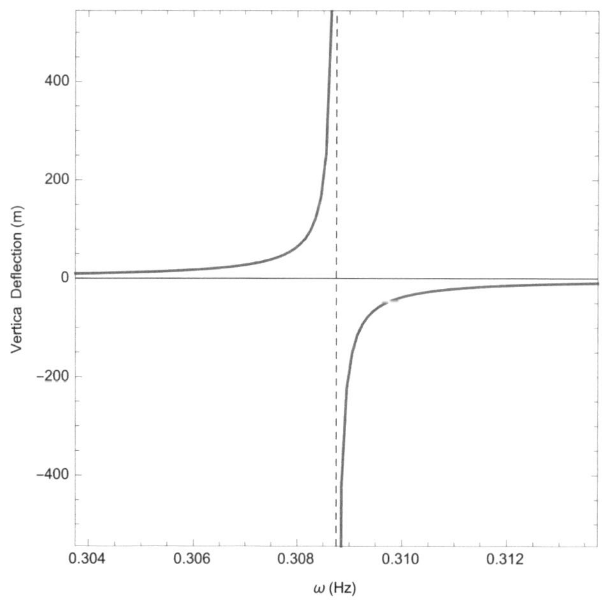

Figure 1.
Large amplitude linear vibrations of the Cosserat body forced to vibrate close to its natural frequency ω_1.

	ω_1, ω_2	ω_3, ω_7	ω_4	ω_5, ω_8	ω_6, ω_9
Plate theory	0.310	17.881	501.13	205.62	338.95
D Cosserat elasticity	0.309	17.763	530.82	211.98	317.87

Table 1.
Comparison of the eigenfrequencies ω_i (Hz) with the exact values of the 3D Cosserat elasticity.

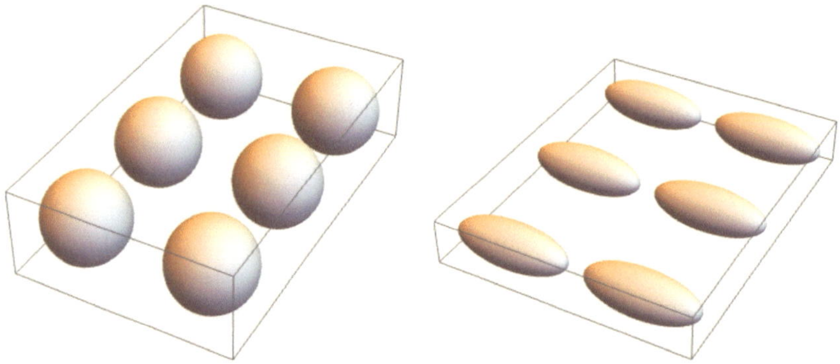

Figure 2.
Ball-shaped micro-elements: $J_x = 0.001, J_y = 0.001, J_z = 0.001$ *(left) and horizontally stretched ellipsoid micro-elements:* $J_x = 0.002, J_y = 0.001, J_z = 0.0001$*(right).*

5.2 Analysis of Cosserat plate vibrations based on the plate theory

We consider a plate $a \times a$ of thickness h with the boundary $G = G_1 \cup G_2$

$$G_1 = \{(x_1, x_2) : x_1 \in \{0, a\}, x_2 \in [0, a]\}$$
$$G_2 = \{(x_1, x_2) : x_2 \in \{0, a\}, x_1 \in [0, a]\}$$

and the following hard simply supported boundary conditions [7]:

$$G_1 : W = 0, W^* = 0, \Psi_2 = 0, \Omega_1^0 = 0, \hat{\Omega}_1^0 = 0, \Omega_3 = 0, \frac{\partial \Psi_1}{\partial n} = 0, \frac{\partial \Omega_2^0}{\partial n} = 0, \frac{\partial \hat{\Omega}_2^0}{\partial n} = 0;$$

$$G_2 : W = 0, W^* = 0, \Psi_1 = 0, \Omega_2^0 = 0, \hat{\Omega}_2^0 = 0, \Omega_3 = 0, \frac{\partial \Psi_2}{\partial n} = 0, \frac{\partial \Omega_1^0}{\partial n} = 0, \frac{\partial \hat{\Omega}_1^0}{\partial n} = 0.$$

Similar to [12] we apply the method of separation of variables for the eigenvalue problem Eq. (85) to solve for the kinematic variables Ψ_α, W, Ω_3, Ω_α^0, W^* and Ω_α^0. The kinematic variables can be further expressed in the following form

$$\Psi_1^{nm} = A_1 \cos\left(\frac{n\pi x_1}{a}\right) \sin\left(\frac{m\pi x_2}{a}\right) \sin(\omega t) + B_1 \sin\left(\frac{n\pi x_1}{a}\right) \cos\left(\frac{m\pi x_2}{a}\right) \sin(\omega t),$$

$$\Psi_2^{nm} = A_2 \sin\left(\frac{n\pi x_1}{a}\right) \cos\left(\frac{m\pi x_2}{a}\right) \sin(\omega t) + B_2 \cos\left(\frac{n\pi x_1}{a}\right) \sin\left(\frac{m\pi x_2}{a}\right) \sin(\omega t),$$

$$W^{nm} = A_3 \sin\left(\frac{n\pi x_1}{a}\right) \sin\left(\frac{m\pi x_2}{a}\right) \sin(\omega t) + B_3 \cos\left(\frac{n\pi x_1}{a}\right) \cos\left(\frac{m\pi x_2}{a}\right) \sin(\omega t),$$

$$\Omega_3^{nm} = A_4 \cos\left(\frac{n\pi x_1}{a}\right) \cos\left(\frac{m\pi x_2}{a}\right) \sin(\omega t) + B_4 \sin\left(\frac{n\pi x_1}{a}\right) \sin\left(\frac{m\pi x_2}{a}\right) \sin(\omega t),$$

$$\Omega_1^{0,nm} = A_5 \sin\left(\frac{n\pi x_1}{a}\right) \cos\left(\frac{m\pi x_2}{a}\right) \sin(\omega t) + B_5 \cos\left(\frac{n\pi x_1}{a}\right) \sin\left(\frac{m\pi x_2}{a}\right) \sin(\omega t),$$

$$\Omega_2^{0,nm} = A_6 \cos\left(\frac{n\pi x_1}{a}\right) \sin\left(\frac{m\pi x_2}{a}\right) \sin(\omega t) + B_6 \sin\left(\frac{n\pi x_1}{a}\right) \cos\left(\frac{m\pi x_2}{a}\right) \sin(\omega t),$$

$$W^{*,nm} = A_7 \sin\left(\frac{n\pi x_1}{a}\right) \sin\left(\frac{m\pi x_2}{a}\right) \sin(\omega t) + B_7 \cos\left(\frac{n\pi x_1}{a}\right) \cos\left(\frac{m\pi x_2}{a}\right) \sin(\omega t),$$

$$\hat{\Omega}_1^{0,nm} = A_8 \sin\left(\frac{n\pi x_1}{a}\right) \cos\left(\frac{m\pi x_2}{a}\right) \sin(\omega t) + B_8 \cos\left(\frac{n\pi x_1}{a}\right) \sin\left(\frac{m\pi x_2}{a}\right) \sin(\omega t),$$

$$\hat{\Omega}_2^{0,nm} = A_9 \cos\left(\frac{n\pi x_1}{a}\right) \sin\left(\frac{m\pi x_2}{a}\right) \sin(\omega t) + B_9 \sin\left(\frac{n\pi x_1}{a}\right) \cos\left(\frac{m\pi x_2}{a}\right) \sin(\omega t),$$

where A_i and B_i are constants.

	ω_1, ω_2	ω_3, ω_7	ω_4	ω_5, ω_8	ω_6, ω_9			
Shape	J_x	J_y	J_z	ω_1, ω_2	ω_3, ω_7	ω_4	ω_5, ω_8	ω_6, ω_9
Ball	0.001	0.001	0.001	17.88	0.31	501.13	205.62	338.95
Vertical ellipsoid	0.001	0.001	0.0001	17.88	0.31	501.13	650.22	338.95
Horizontal ellipsoid	0.0001	0.001	0.001	17.88	0.31	1363.01	205.62	394.08

Table 2.
Eigenfrequencies ω_i^{11} *(Hz) for different shapes of micro-elements.*

We solve an eigenvalue problem by substituting these expressions into the system of Eq. (85). The obtained nine sequences of positive eigenfrequencies ω_i^{nm} are associated with the rotation of the middle plane (ω_1^{nm} and ω_2^{nm}), flexural motion and its transverse variation (ω_3^{nm} and ω_7^{nm}), micro rotatory inertia (ω_4^{nm}, ω_5^{nm} and ω_6^{nm}) and its transverse variation (ω_8^{nm} and ω_9^{nm}) [12].

We perform all our numerical simulations for $a = 3.0$ m and $h = 0.1$ m. We consider different forms of micro elements: ball-shaped elements, horizontally and vertically stretched ellipsoids (see **Figure 2**). For simplicity we will use the notation ω_i for the first elements ω_i^{11} of the sequences ω_i^{nm}. The results of the computations are given in the **Table 2**. The shape of the micro-elements does not effect the natural macro frequencies ω_1 and ω_2 associated with the rotation of the middle plane and ω_3 and ω_7 associated with the flexural motion and its transverse variation. The ellipsoid elements have higher micro frequencies associated with the micro rotatory inertia (ω_4, ω_5 and ω_6) and its transverse variation (ω_8 and ω_9), than the ball-shaped elements.

Let J_x, J_y and J_z be the principal moments of inertia of the microelements corresponding to the principal axes of their rotation. We assume that the quantities J_x, J_y and J_z are constant throughout the plate B_0. If the microelements are rotated around the z-axis by the angle θ the rotatory inertia tensor $\mathbf{J}$ can be expressed as

$$\mathbf{J} = \begin{pmatrix} J_x \cos^2\theta + J_y \sin^2\theta & \left(J_x - J_y\right) \sin 2\theta & 0 \\ \left(J_x - J_y\right) \sin 2\theta & J_x \sin^2\theta + J_y \cos^2\theta & 0 \\ 0 & 0 & J_z \end{pmatrix} \tag{103}$$

Angle θ	ω_1	ω_2	ω_3	ω_7	ω_4	ω_5	ω_8	ω_6	ω_9
0°	17.88	17.88	0.31	0.31	650.221	265.37	265.37	450.61	450.61
10°	17.88	17.88	0.31	0.31	650.221	255.59	279.40	429.89	469.93
20°	17.88	17.88	0.31	0.31	650.221	247.75	295.33	406.70	484.79
30°	17.88	17.88	0.31	0.31	650.221	242.57	313.65	382.94	495.14
40°	17.88	17.88	0.31	0.31	650.221	239.99	333.10	360.57	500.46
45°	17.88	17.88	0.31	0.31	650.221	239.68	338.95	354.35	501.13
50°	17.88	17.88	0.31	0.31	650.221	239.99	333.10	360.57	500.46
60°	17.88	17.88	0.31	0.31	650.221	242.57	313.65	382.94	495.14
70°	17.88	17.88	0.31	0.31	650.221	247.75	295.33	406.70	484.79
80°	17.88	17.88	0.31	0.31	650.221	255.59	279.40	429.89	469.93
90°	17.88	17.88	0.31	0.31	650.221	265.37	265.37	450.61	450.61

Table 3.
Eigenfrequencies ω_i^{11} *(Hz) for different angles of rotation of horizontal ellipsoid micro-elements.*

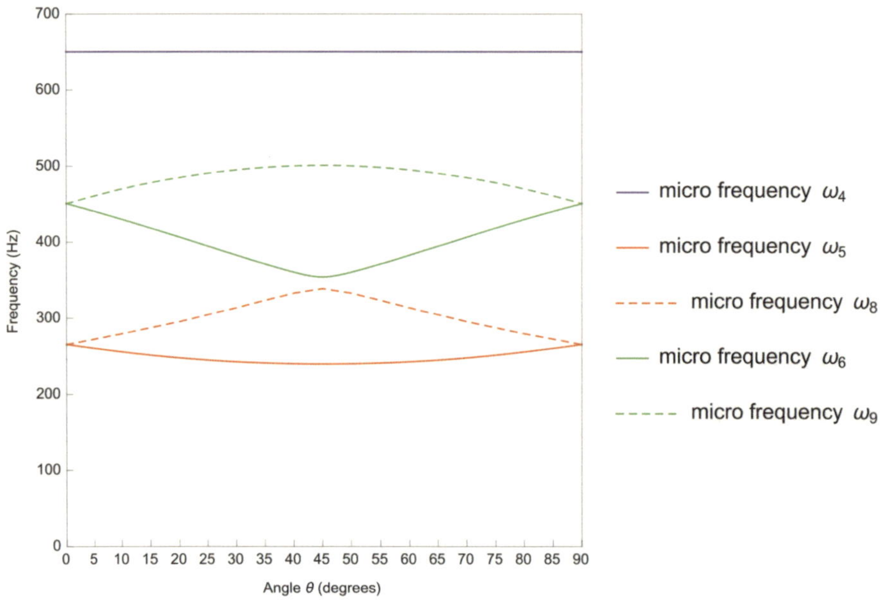

Figure 3.
Micro frequencies ω_4, ω_5, ω_8, ω_6 and ω_9.

The eigenfrequencies for different angles of microrotation of the microelements are given in the **Table 3** and the **Figure 3**. The rotatory inertia principle moments used are $J_x = 0.002, J_y = 0.001, J_z = 0.0001$, which represent a horizontally stretched ellipsoid microelement. The case when the microelements are not aligned with the edges of the plate the model predicts some additional natural frequencies related with the microstructure of the material.

6. Conclusions

In this chapter, we presented a mathematical model of Cosserat plate vibrations. The dynamic model of the plates has been developed as a dynamic extension of the Reissner plate theory. The equations has been presented in both tensorial and the matrix forms. We also described the validation of the model, which is based on the comparison with the three-dimensional Cosserat elastodynamics exact solutions. Based on the presented results of the computer simulations we were able to detect and classify the additional high resonance frequencies of a plate. We have shown that the frequencies depend on the shape and orientation of microelements (ball-shaped elements, horizontally and vertically stretched ellipsoids) incorporated into the Cosserat plates. We also have been able to identify that micro frequencies associated with the micro rotatory inertia and its transverse variation of the ellipsoid elements have higher micro frequencies than the ball-shaped elements. We also showed the dependence of the eigenfrequencies on the angles of rotation of the horizontal ellipsoid micro-elements. These results can be used to identify the characteristics of the plate micro-elements.

Appendix A: conventions and notations

A.1 Conventions

We use the following notation convention:

1. the values of the Latin subindex i take values in the set $\{1, 2, 3\}$

2. the values of the Greek indices α and β take values in the set $\{1, 2\}$

3. the Einstein summation notation is used throughout the chapter

A.2 Notations

$\{x_i\}$	artesian coordinates
P	Cosserat thin plate
h	plate thickness
μ, λ	Lamé parameters
$\alpha, \beta, \gamma, \varepsilon$	Cosserat elasticity parameters
ρ	material density
J_{ji} or $\mathbf{J}$	rotatory inertia
σ_{ji} or $\boldsymbol{\sigma}$	the stress tensor
μ_{ji} or $\boldsymbol{\mu}$	the couple stress tensor
γ_{ji} or $\boldsymbol{\gamma}$	strain tensor
χ_{ji} or $\boldsymbol{\chi}$	bend-twist tensor
u_i or $\mathbf{u}$	displacement vector
ϕ_i or $\boldsymbol{\phi}$	microrotation vector
p_i or $\mathbf{p}$	linear momentum
q_i or $\mathbf{q}$	angular momentum
ε_{ijk}	Levi-Civita tensor
U_C	strain stored energy
U_K	stress energy
T_C	stored kinetic energy
T_W	work of inertia forces
$\mathcal{S}$	Cosserat plate stress set
$\mathcal{U}$	Cosserat plate displacement set
$\mathcal{E}$	Cosserat plate strain set
η	splitting parameter
p	pressure
ω	natural frequency of plate vibration
θ	angle of microelement orientation
$M_{\alpha\beta}$	bending and twisting moments
Q_α	shear forces
$Q_\alpha^*, \hat{Q}_\alpha$	transverse shear forces
$R_{\alpha\beta}$	micropolar bending moments
$R_{\alpha\beta}^*$	micropolar twisting moments
S_α^*	micropolar couple moments

Ψ_α	rotations of the middle plane around x_α axis
W, W^*	vertical deflections of the middle plate
Ω_α^0	microrotations in the middle plate around x_α axis
Ω_3	rate of change of the microrotation

Author details

Lev Steinberg[1]* and Roman Kvasov[2]

1 Department of Mathematical Sciences, University of Puerto Rico at Mayagüez, Mayagüez, Puerto Rico, USA

2 Department of Mathematics, University of Puerto Rico at Aguadilla, Aguadilla, Puerto Rico, USA

*Address all correspondence to: lev.steinberg@upr.edu

References

[1] Cosserat E, Cosserat F. Theorie des corps deformables [Theory of Deformable Bodies]. Paris: A. Hermann et fils; 1909

[2] Green A, Naghdi P. The linear theory of an elastic Cosserat plate. Mathematical Proceedings of the Cambridge Philosophical Society. 1966; **63**:537-550

[3] Eringen A. Theory of micropolar plates. Journal of Applied Mathematics and Physics. 1967;**18**:12-31

[4] Altenbach H, Eremeyev V. On the theories of plates based on the Cosserat approach. Advances in Applied Mathematics and Mechanics. 2010;**21**: 73-92

[5] Steinberg L. Deformation of micropolar plates of moderate thickness. International Journal of Applied Mathematics and Mechanics. 2010; **6**(17):1-24

[6] Kvasov R, Steinberg L. Numerical modeling of bending of Cosserat elastic plates. In: Proceedings of the 5th Computing Alliance of Hispanic-Serving Institutions; 2011. pp. 67 70

[7] Steinberg L, Kvasov R. Enhanced mathematical model for Cosserat plate bending. Thin-Walled Structures. 2013; **63**:51-62

[8] Reissner E. On the theory of elastic plates. Journal of Mathematics and Physics. 1944;**23**:184-191

[9] Reissner E. The effect of transverse shear deformation on the bending of elastic plates. Journal of Applied Mechanics. 1945:69-77

[10] Kvasov R, Steinberg L. Numerical modeling of bending of micropolar plates. Thin-Walled Structures. 2013;**69**: 67-78

[11] Kvasov R, Steinberg L. Modeling of size effects in bending of perforated Cosserat plates. Modelling and Simulation in Engineering. 2017;**2017**:1-19

[12] Steinberg L, Kvasov R. Analytical modeling of vibration of micropolar plates. Applied Mathematics. 2015;**6**: 817-836

[13] Nowacki W. Theory of Asymmetric Elasticity. Oxford, New York: Pergamon Press; 1986

[14] Gurtin M. The linear theory of elasticity. In: Truesdell C, editor. Handbuch der Physik. Vol. VIa/2. Berlin: Springer-Verlag; 1972. pp. 1-296

[15] Lakes R. Experimental methods for study of Cosserat elastic solids and other generalized elastic continua. In: Mühlhaus H, editor. Continuum Models for Materials with Microstructures. New York: Wiley J; 1995. pp. 1-22

[16] Singh S. Blowing agents for polyurethane foams. Rapra Review Report. 2002;**12**(10)

Chapter 6

On the Controlling of Multi-Legged Walking Robots on Stable and Unstable Ground

Dariusz Grzelczyk and Jan Awrejcewicz

Abstract

In this chapter, we developed and investigated numerically a general kinematic model of a multi-legged hybrid robot equipped with a crab-like and/or mammal-like legs. To drive the robot's limbs, a novel generator of gait was employed and tested. The simulation model developed in Mathematica is suitable for virtual study and visualization of the locomotion process. In contrast to our previous papers, in this study we focused especially on precise control of the position of the robot during walking in different directions. In our study we were able to simultaneously control all six spatial degrees of freedom of the robot's body, as well as all the robot's legs. Therefore, the investigated robot can be considered and used as a fully controlled walking Stewart platform. What is more, the used algorithm can also be successfully employed to coordinate and control all limbs of the robot on unstable or vibrating ground. As an example, it can be used to stabilize spatial position of the robot when the supporting ground becomes vibrating or unstable, and it will keep the robot stable and prevent it from falling over. Eventually, the developed simulation algorithms can be relatively simply adopted to control real constructions of different multi-legged robots.

Keywords: legged motion, robot gait, robot control, robot kinematics, robot dynamics, robot stability

1. Introduction

Different kinds of multi-legged robots are extensively investigated in the literature, since they have great importance in engineering applications [1]. In general, each legged robot can be considered as a mechatronic system with limbs connected to the main body (i.e., the robot's trunk). In such a construction of the walking machine, the robot's legs are responsible for acting as a support frame and play a key role when it comes to the locomotion process. Namely, when the limbs of the robot are controlled with a degree of autonomy, it can move within its environment and perform the planned tasks [2]. For instance, the legged machines are able to reproduce animal or human movements and substitute humans in various activities [3]. An interesting state of the art regarding the most popular hexapod robots can be found in numerous review papers. Among numerous multi-legged robots met in the literature, we can distinguish especially robots with leg structure inspired based on the anatomy of insects or mammals. Large numbers of legs with different kinematic structures are useful to overcome complex obstacles met in nature without losing stability and increase stability of the robot.

About 50 percent of the Earth's land surface is not adapted for wheeled machines. However, it can be accessed by humans and animals that can walk in those difficult and irregular terrains. Therefore, nowadays, there is a lot of interest in different kinds of robots that use legged motion inspired by walking animals found in nature. This type of motion allows to overcome obstacles, move in an omnidirectional manner, and access uneven environments, and it is fault tolerant, in comparison to crawler or wheeled vehicles. Unfortunately, legged robots are challenging in terms of controlling their locomotion. However, on the other hand, they can be used in terrains, where wheeled machines cannot perform their tasks. It is because legged robots can overcome obstacles of heights equal to the height of their limbs, while the wheeled robots can overcome obstacles of heights up to the half of the wheel radius. On the contrary to wheeled robots, where the contact between the robot's wheels and the ground usually has a continuous character, in the case of legged robots, the contact between the feet and the ground usually takes a form of contact points. It is known that a large amplitude of the ground reaction force acting on the robot's feet has a negative impact on the dynamics of the whole robot. Therefore, these forces should be minimized, if possible. This is why one of the considered issues is also related to the problem of minimization of the ground reaction forces acting on the robot's legs.

A literature review summarized in our previous papers [4–11] indicate that investigations of different types of walking machines are still challenging and focus the attention of numerous researchers. The control possibilities can be especially helpful in a natural environment of the robot, when it comes to both the navigation and obstacle avoidance. That is why, in this chapter we developed a general full parametric simulation model of a hybrid walking robot, i.e., the robot which has different numbers of the legs inspired biologically by insects, reptiles, or mammals. To drive the legs of the abovementioned robot, we employed a novel gait generator, firstly introduced in our previous paper [11]. The mentioned approach can be treated as neural networks that generate rhythmic outputs in the absence of rhythmic input. Such kinds of rhythmic motions have been found by biologists in different biological periodic processes, including swimming, running, breathing, flying, chewing, or walking. Moreover, we also used own algorithm, which is suitable for a smooth transition between different gait phases, i.e., initial, rhythmic, and terminal phases [10]. The main goal of the abovementioned studies was to obtain better both kinematic and dynamical parameters of the investigated machines during the locomotion process, which finally lead to an increase in the stability of the robot and its control possibilities. Eventually, we considered the problem of controlling the direction of the movement of the robot and control all six spatial degrees of freedom of the robot's body, i.e., three deviations and three rotations along and around three different axes, respectively. It can be useful for the walking robot and to control all robot's legs on a planar, unstable, and vibrating ground, since both the navigation and obstacle avoidance matters are particularly important for the locomotion of legged robots in the natural environment.

2. A brief summary of our chosen previous studies related to walking robots

Our numerical investigations related to the kinematical analysis and control of the six-legged walking robot have been initiated in the paper [4]. To control the movement of the tip of the robot leg, a nonlinear mechanical oscillator describing stick–slip mechanical vibrations has been proposed and applied as the central pattern generator (CPG). By using also three well-known mechanical oscillators (i.e., Hopf oscillator, Van der Pol oscillator, and Rayleigh oscillator), the advantages of the proposed CPG model have been illustrated and discussed. Time histories of the articulated variables in all joints of the robot's leg and its configurations during walking indicated some analogies between the characteristics of the simulated walking robot and animals found in nature. Eventually, the results obtained in that paper indicated that, from the point of view of the energy demand, the proposed model of CPG based on the stick–slip vibrations can be more efficient than other previous CPG models found in the literature.

In the next paper [5], the above-presented study has been continued regarding power consumption experimental analysis with the use of the constructed hexapod robot controlled by previously introduced CPGs. The power consumption of the constructed robot has been calculated based on the current consumption in the servo motors, which drive the robot's legs. The performed experimental investigations showed different energy demand, depending on the CPG model driving the robot's legs. By comparing the total energy demand of all servomechanisms installed in the robot for the same locomotion conditions, it has been proven that the proposed CPG model is characterized by the lowest energy demand.

The study [6] introduced dynamical modeling of the previously investigated hexapod robot walking with the tripod gait on a planar and hard ground. This type of gait is commonly used by six-legged real insects as well as six-legged walking machines met in engineering applications. Using the inverse dynamic approach and some simplifications resulting from the symmetry of the robot, vertical components of the ground reaction forces acting on the robot's legs have been estimated numerically. As a result, different time histories of the contact forces and overloads acting on the robot's legs have been detected, depending on the CPG model driving the robot's legs. Finally, the obtained information can be used for further strength analysis of the robot's legs to ensure trouble-free use and an extension of life and operational time of the robot.

On the contrary to the abovementioned papers, in Ref. [7], a new model of an octopod (eight-legged) robot has been developed and numerically simulated in order to investigate crucial kinematic parameters during locomotion with the tetrapod gait, i.e., similar gait as the tripod gait used in hexapod robots. Another well-known CPG model has also been used (i.e., CPG model constructed based on the Toda-Rayleigh lattice). However, the main novelty of that article is a proposition of a new model of gait generator constructed based on the sine function, further referred as a SINE generator. The proposed model of gait is relatively simple in comparison with other control methods found in the literature and can be especially useful when the robot walks on planar/regular surfaces. Some advantages of the proposed SINE generator have been outlined, including the lack of the acceleration and deceleration of the robot gravity center. These advantages have a positive influence on the robot's dynamics, especially when it comes to the minimization of the ground reaction forces acting on the robot's legs and low energy consumption of the robot during walking.

Numerical investigations of the octopod robot and a novel gait generator introduced in the paper [7] have been significantly extended in Ref. [8]. Aside from computing crucial kinematic parameters of the robot, also time histories of the robot's dynamic locomotion parameters have been illustrated and discussed. The obtained dynamic parameters confirmed some advantages of the proposed SINE generator previously presented in paper [7]. Concluding, it should be mentioned that the proposed SINE generator is relatively simple in comparison with other control methods found in the literature. Especially, solving of nonlinear differential equations is not required in this model, on the contrary to all previously presented CPGs. The proposed model of the gait does not produce the unnecessary fluctuations in the center of gravity of the robot or variations in the acceleration/deceleration in the direction of the robot movement. This is why the proposed generator appears to be more efficient with respect to the energy demand than other CPGs investigated in that paper.

In the next paper [9], another kinematic/dynamic simulation model of the hexapod robot has been developed and implemented in Mathematica. On the contrary to previous crab-like hexapod/octopod robots, another kinematic model of a single robot's leg has been used. Also in that study, advantages of the proposed gait generator have been clearly emphasized, especially with regard to the kinematic parameters (displacement and velocity of the whole robot) as well as dynamic parameters (ground reaction forces and overloads acting on the robot legs). In addition, the problem of trajectory planning of the position of the robot body during the walking process has been considered. In the developed simulation model, the appropriate fluctuations of the robot in the vertical direction during the walking process can be controlled precisely. Such a control approach can be useful in the natural environment, for instance, when the robot is walking under or over obstacles.

In paper [10], novel kinematic and dynamic models of a mammal-like octopod robot have been developed and numerically investigated. In order to control the robot's legs, also new simple gait generators constructed based on a sine function have been employed and tested. The proposed models are relatively simple in comparison to other control methods found in the literature. It has been shown that the used gait generators can be useful to obtain better both kinematic and dynamic properties of motion of the whole robot. The advantages of the proposed models have been clearly emphasized, especially with regard to the crucial kinematic and dynamic parameters. Another novelty of this paper is a proposition of one model to control the initial, regular, and terminal phases of the robot gait without the need for generating additional control signals. The developed simulation model of the robot allows also to precisely control robot's vertical position. As a result, better stability of the whole robot during walking and performing the planned tasks, also on terrains characterized by a relatively low friction coefficient between the ground and the robot feet, has been detected.

The study presented in paper [11] is a significant extension of the control methods introduced in the paper [10]. To drive the robot's legs, also another nonlinear oscillator (i.e., hybrid Van der Pol-Rayleigh oscillator) has been used as a central pattern generator. Moreover, also a new, relatively simple, and efficient model has been proposed and tested. The proposed model of the gait generator is useful to obtain better both kinematic and dynamic parameters of motion of the robot walking in different directions. By changing the length and the height of a single step of the robot, the initial, rhythmic, and terminal phases of the robot gait, as in the previous paper [10], have been introduced. Using the simulation model developed in Mathematica, displacement, velocity and acceleration of the center of the robot's body, fluctuations in the zero moment point of the robot, and the ground reaction forces acting on the feet of the robot have been computed, reported, and discussed. The obtained results showed advantages of the introduced model of robot's gaits regarding fluctuations in the robot's body, the minimum value of dynamic stability margin, as well as the minimum value of a friction coefficient which is necessary to avoid slipping between the ground and the robot's feet during the walking process. Moreover, the proposed model does not produce unnecessary fluctuations in the velocity both in the vertical and horizontal (i.e., movement and lateral) directions of the robot. As a result, it has also a positive impact on the dynamical parameters of the robot. In addition, as some of the previously discussed models, the proposed gait generator is relatively simple in comparison to the other three tested CPGs constructed on the basis of the nonlinear oscillators. Finally, the employed model of gait also allowed to precisely control the vertical position of the robot during walking in different directions.

Also recently, various control strategies of the walking machines were usually tested by other researches using different commercial software such as MATLAB [2, 12, 13], ADAMS [12, 14–16], or the Open Dynamics Engine [17, 18]. It shows that investigations of walking robots are still challenging for researches and focus their attention. Therefore, as it was shortly mentioned above, in this chapter we developed a general full parametric simulation model of a hybrid walking robot, i.e., the robot which can have different numbers of legs inspired biologically by insects, reptiles, or mammals. To drive the legs of the abovementioned robot, we employed central pattern generator, firstly introduced in our previous paper [11]. Moreover, we also used own algorithm, which is suitable for a smooth transition between different gait phases, i.e., initial, rhythmic, and terminal phases [10]. Eventually, we considered the problem of controlling the direction of the movement of the robot and control all six spatial degrees of freedom of the robot's body (three deviations and three rotations along and around three different axes, respectively), as well as control all robot's legs on planar, vibrating, and unstable ground. These control possibilities can be especially helpful in a natural environment of the robot, when it comes to both the navigation and obstacle avoidance.

3. Prototypes of different walking robots

Advances computer modeling can be useful to increase the efficiency of designed machines, by introducing required modifications to the mechanical construction, depending on the results obtained during virtual tests. Currently, numerous computer programs, such CATIA, SimDesigner, ADAMS, or SolidWorks software, support researchers in virtual prototyping of machines, including walking robots, for instance, see papers [19–21]. In our study we also designed different walking robots, with different numbers of limbs and their different structures. All our designs presented in **Figure 1** were created in Inventor Professional 2019 software. In general, each robot consists of the main body part (trunk) and four

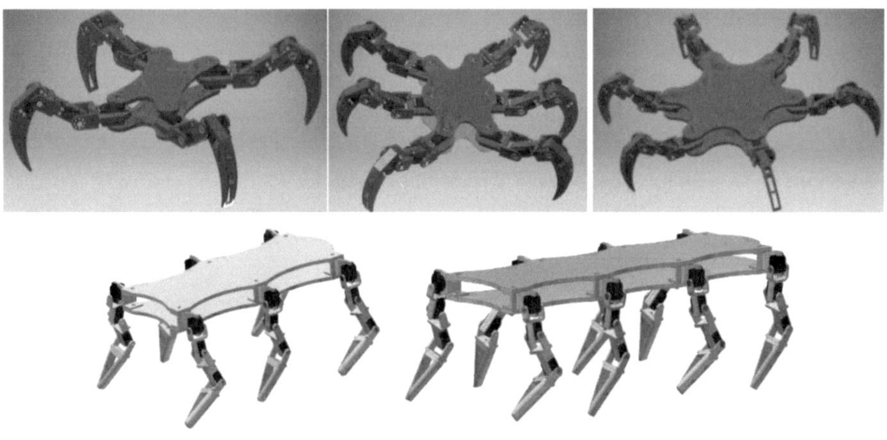

Figure 1.
CAD models of different bio-inspired walking robots created in inventor professional 2019 software.

(quadruped), six (hexapod), or eight (octopod) legs with crab-like or mammal-like structures. Following many previous researches, during modeling of all robot's legs, we were inspired by the morphology of limbs of typical insects or mammals, reduced to three main segments [2, 12, 22–27]. To avoid mutual collisions between the legs, in all presented designs, the robot's legs are symmetrically distributed on two sides of the robot's body. By adopting the initial positions of individual leg's joints, different leg's configurations, which correspond to configurations of lower limbs of different animals met in nature, can be implemented [15]. In all designs presented in **Figure 1**, each joint of the robot's leg is actuated by Power HD AR-1201MG servomotor controlled by the pulse-width modulation (PWM) technique. This servomotor has inner closed loop feedback of the position control and a proportional controller.

Both kinematic and dynamic models of crab-like or mammal-like robots were detail presented in our previous papers [8, 10, 11]. On the contrary to the abovementioned references, in this study we focused only on novel mathematical formulations regarding full control of the robot on a planar, irregular, unstable, or vibrating ground, which have not been presented before. To investigate the possibility of the robot's control, we have developed a general, three-dimensional, fully parametric simulation model of a walking robot created in the Mathematica software. The created simulation model allowed us to visualize behavior of the investigated robot and control the correctness of the obtained results. Especially, we were able to control the spatial position of individual elements of the robot and the configurations of the legs. It also allowed us to better understand the obtained simulation results. Through the appropriate choosing of the parameters and configurations of the limbs, we are able to quickly generate various types of walking robots.

4. Kinematics of the robot

In one of the recent papers [11], we considered and investigated the problem of planning the trajectory of the robot body in the vertical direction, during walking on a planar plane in any (i.e., forward, lateral, oblique, etc.) direction. On the contrary to the mentioned paper, in this study we extended significantly the control possibility of the investigated robot, including full control of all six spatial degrees of freedom of the robot's body (i.e., three deviations and three rotations along and around three different axes, respectively), as well as control of all robot's legs on planar, non-regular, unstable, and vibrating ground. As it was mentioned before, this control possibility can be really helpful, when it comes to both the navigation and obstacle avoidance in natural environment of the robot. As a result, the presented robot can be treated as a fully controlled walking Stewart platform. The created simulation models allow to visualize the investigated robots during walking and control the correctness of the obtained results. Especially, it was able to control the spatial position of individual elements of the robot and the configurations of the legs. Finally, it allows to understand the obtained simulation results better.

Figure 2 shows exemplary simulation model of a hybrid octopod robot. Four of its legs (two front and two hind) have mammal-like structures, whereas four middle legs have crab-like structures. In addition, for more general considerations, we used different initial configurations of the mammal-like legs for front and hind legs. Model of this hybrid octopod robot standing on a planar surface is presented in **Figure 2**.

The robot is embedded in a global coordinate system $Oxyz$. In addition, other local coordinate systems are fixed to the robot. The coordinate system $O'x'y'z'$ is associated with the center of the robot's body. In turn, other local coordinate

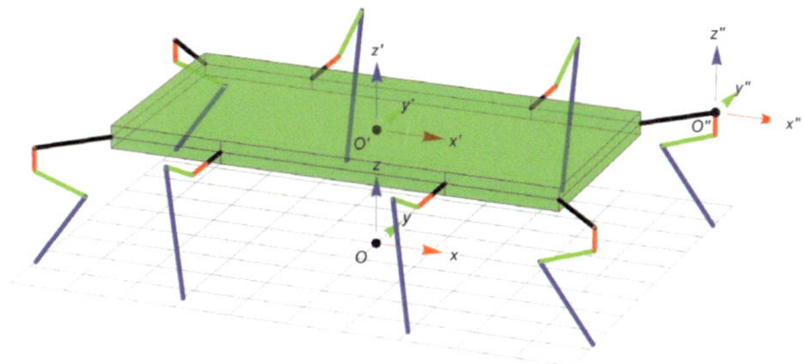

Figure 2.
Simulation model of a hybrid octopod walking robot.

systems are associated with the robot's legs. In **Figure 2** we presented only one local coordinate system $O''x''y''z''$ fixed to the leg L1.

In general, we consider irregular and unstable ground, where a height (z- component) for the coordinates x and y for time t has the following general form:

$$z = z_G(x, y, t). \tag{1}$$

The vectors $\mathbf{r}_G(t)$ describing positions of all points on the ground in the global coordinate system $Oxyz$ have the form

$$\mathbf{r}_G(t) = \mathbf{R}_G(\alpha_G(t), \beta_G(t), \gamma_G(t)) \cdot \mathbf{r}_G^{init} + \Delta\mathbf{r}_G(t), \tag{2}$$

where $\Delta\mathbf{r}_G(t) = [\Delta x_G(t), \Delta y_G(t), \Delta z_G(t)]^{\mathrm{T}}$, $\mathbf{r}_G^{init}$ are vectors describing positions of points of the ground in the initial time, i.e., for the angles $\alpha_G(t) = 0$, $\beta_G(t) = 0$, $\gamma_G(t) = 0$ and linear deviations $\Delta x_G(t) = 0$, $\Delta y_G(t) = 0$, and $\Delta z_G(t) = 0$,

$$\mathbf{R}_G(\alpha_G(t), \beta_G(t), \gamma_G(t)) = \mathbf{R}_G^{(z)}(\gamma_G(t)) \cdot \mathbf{R}_G^{(y)}(\beta_G(t)) \cdot \mathbf{R}_G^{(x)}(\alpha_G(t)) \tag{3}$$

is the rotation matrix, whereas

$$\mathbf{R}_G^{(x)}(\alpha_G(t)) = \begin{bmatrix} 1 & 0 & 0 \\ 0 & \cos\alpha_G(t) & -\sin\alpha_G(t) \\ 0 & \sin\alpha_G(t) & \cos\alpha_G(t) \end{bmatrix}, \tag{4}$$

$$\mathbf{R}_G^{(y)}(\beta_G(t)) = \begin{bmatrix} \cos\beta_G(t) & 0 & \sin\beta_G(t) \\ 0 & 1 & 0 \\ -\sin\beta_G(t) & 0 & \cos\beta_G(t) \end{bmatrix}, \tag{5}$$

$$\mathbf{R}_G^{(z)}(\gamma_G(t)) = \begin{bmatrix} \cos\gamma_G(t) & -\sin\gamma_G(t) & 0 \\ \sin\gamma_G(t) & \cos\gamma_G(t) & 0 \\ 0 & 0 & 1 \end{bmatrix} \tag{6}$$

are elementary rotation matrices of the ground, respectively. All vectors $\mathbf{r}_R(t)$ of the robot in the global coordinate system $Oxyz$ are expressed by

$$\mathbf{r}_R(t) = \mathbf{R}_R(\alpha_R(t), \beta_R(t), \gamma_R(t)) \cdot \mathbf{r}_R^{init} + \Delta\mathbf{r}_R(t), \tag{7}$$

where $\Delta\mathbf{r}_R(t) = [\Delta x_R(t), \Delta y_R(t), \Delta z_R(t)]^{\mathrm{T}}$, $\mathbf{r}_R^{init}$ are vectors describing positions of points of the robot in the initial time, i.e., for the angles $\alpha_R(t) = 0$, $\beta_R(t) = 0$, and $\gamma_R(t) = 0$ and linear deviations $\Delta x_R(t) = 0$, $\Delta y_R(t) = 0$, and $\Delta z_R(t) = 0$,

$$\mathbf{R}_R(\alpha_R(t),\beta_R(t),\gamma_R(t)) = \mathbf{R}_R^{(z)}(\gamma_R(t)) \cdot \mathbf{R}_R^{(y)}(\beta_R(t)) \cdot \mathbf{R}_R^{(x)}(\alpha_R(t)) \tag{8}$$

is the rotation matrix, whereas

$$\mathbf{R}_R^{(x)}(\alpha_R(t)) = \begin{bmatrix} 1 & 0 & 0 \\ 0 & \cos\alpha_R(t) & -\sin\alpha_R(t) \\ 0 & \sin\alpha_R(t) & \cos\alpha_R(t) \end{bmatrix}, \tag{9}$$

$$\mathbf{R}_R^{(y)}(\beta_R(t)) = \begin{bmatrix} \cos\beta_R(t) & 0 & \sin\beta_R(t) \\ 0 & 1 & 0 \\ -\sin\beta_R(t) & 0 & \cos\beta_R(t) \end{bmatrix}, \tag{10}$$

$$\mathbf{R}_R^{(z)}(\gamma_R(t)) = \begin{bmatrix} \cos\gamma_R(t) & -\sin\gamma_R(t) & 0 \\ \sin\gamma_R(t) & \cos\gamma_R(t) & 0 \\ 0 & 0 & 1 \end{bmatrix}. \tag{11}$$

The determination of the coordinates describing the positions of the feet of all the robot's legs requires consideration of the vectors $\mathbf{r}_G(t)$ in contact points between the robot's legs and the ground, as well as the inverse matrix $\mathbf{R}_R^{-1}$. Next, the appropriate articulated variables for all joints of the robot's legs can be determined.

5. Simulation results

5.1 Control of the robot walking on a planar, stable, and not vibrating surface

First, we considered the robot walking at an oblique direction on a planar, stable, and not vibrating ground (i.e., $z_G(x,y,t) = 0$, $\Delta x_G(t) = \Delta y_G(t) = \Delta z_G(t) = 0$, and $\alpha_G(t) = \beta_G(t) = \gamma_G(t) = 0$). In the presented simulations, we considered initial, regular, and terminal phases of gait [10]. In turn, we applied non-zero excitations of all six degrees of freedom of the robot's trunk, i.e. $\Delta x_R(t) \neq 0$, $\Delta y_R(t) \neq 0$, $\Delta z_R(t) \neq 0$, $\alpha_R(t) \neq 0$, $\beta_R(t) \neq 0$, $\gamma_R(t) \neq 0$. Configurations of both the robot's body and all its legs captured in regular time intervals are presented in **Figure 3**. In the considered case, the robot moves in an oblique direction with respect to the direction determined by the robot's (x'-) axis at the initial time. Changing the position of the whole robot is realized by the appropriate control of individual joints of the robot's legs. In turn, **Figure 4** presents fluctuations in displacements and velocities of the robot, both in the forward (x-) and lateral (y-) direction. As can be seen, the presented time histories can be divided into three ranges, which correspond to the initial, regular, and terminal phases of the robot locomotion, respectively. Regardless of the $\Delta x_R(t)$, $\Delta y_R(t)$, $\Delta z_R(t)$, $\alpha_R(t)$, $\beta_R(t)$, $\gamma_R(t)$ controlling all six degrees of freedom of the robot's trunk, both x- and y- components of the robot's speed change linearly. Namely, in the initial phase of the locomotion process, both of these speeds increase linearly from zero to the maximum values. Then, in the rhythmic phase of gait, these speeds are constant over time. In turn, in the terminal phase of the locomotion process, these speeds decrease linearly from maximum values to zero. In the same time, linear deviations, as well as rotations of the robot's body in the global coordinate system, are accurately reflected based on the functions $\Delta x_R(t)$, $\Delta y_R(t)$, $\Delta z_R(t)$, and $\alpha_R(t)$, $\beta_R(t)$, $\gamma_R(t)$, respectively. These simulations show that the considered robot can be used as a fully controlled walking Stewart platform. As a result, we solved the problem of controlling both the direction of the movement of

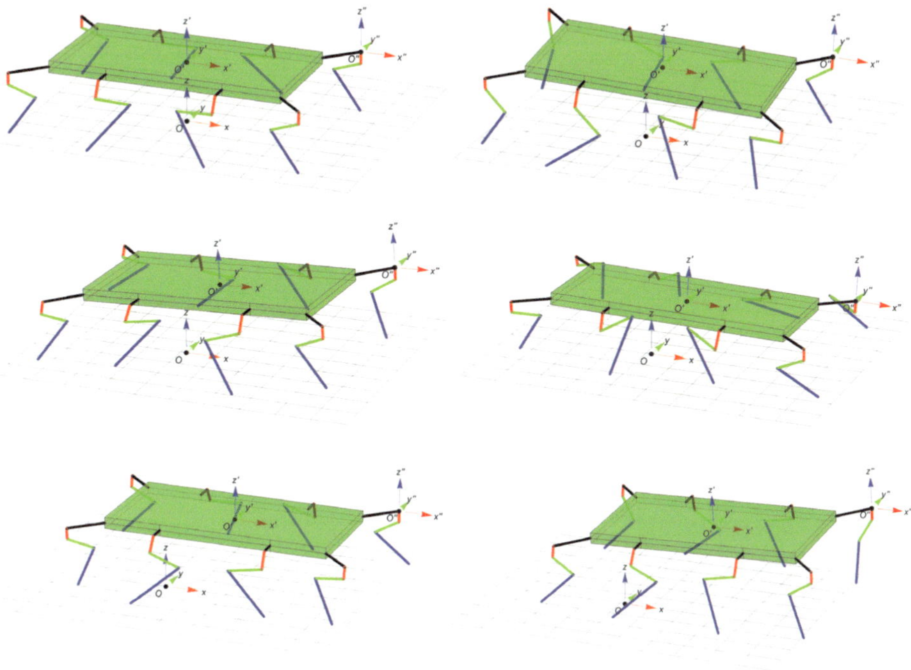

Figure 3.
The chosen configurations of the investigated octopod robot walking on a planar surface, controlled independently by deviations $\Delta x_R(t)$, $\Delta y_R(t)$, $\Delta z_R(t)$ and rotations $\alpha_R(t)$, $\beta_R(t)$, $\gamma_R(t)$ of the robot's trunk.

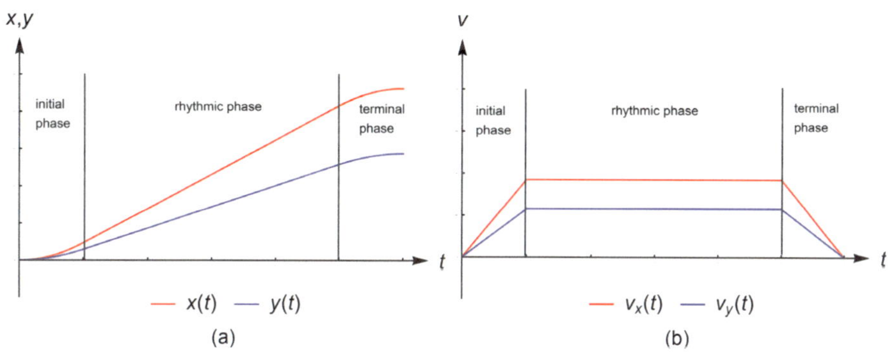

Figure 4.
Time histories of the displacements $x(t)$, $y(t)$ of the robot's center (a) and velocities $v_x(t)$, $v_y(t)$ of the robot's center (b) in the forward and lateral directions.

the robot and the control of all six spatial degrees of freedom of the robot's body, namely, independent controlling of three deviations and three rotations along and around three different axes, respectively. As it was mentioned above, these control possibilities can be useful in the natural environment of the robot, where the navigation and obstacle avoidance are especially important.

5.2 Control of the robot standing on not vibrating and unstable ground

In this subsection we considered the problem of control of the robot on not vibrating but unstable ground. **Figure 5** presents configurations of the robot standing on unstable ground (i.e., $z_G(x, y, t) \neq 0$), captured in regular time intervals. To better illustrate the process of controlling individual legs of the robot on unstable

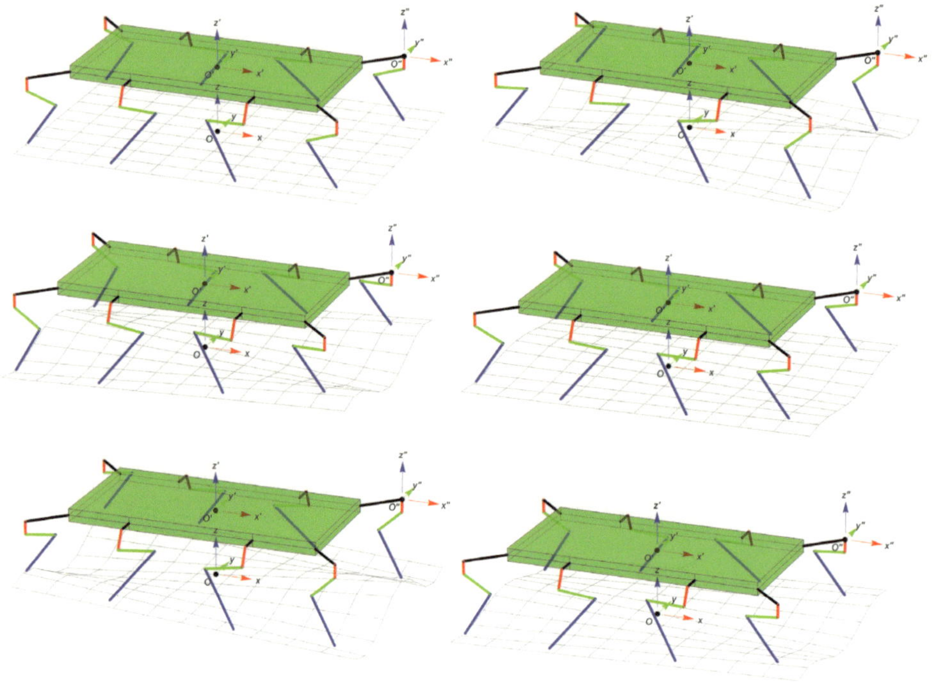

Figure 5.
Configurations of the investigated hybrid octopod robot on unstable ground, captured in regular time intervals.

ground, we assumed that $\Delta x_R(t) = \Delta y_R(t) = \Delta z_R(t) = 0$ and $\alpha_R(t) = \beta_R(t) = \gamma_R(t) = 0$. As a result, we can observe the process of stabilizing the linear and angular positions of the robot, when the supporting ground becomes unstable. As we can see, at any time, the robot is supported by all eight legs, through the appropriate changing of the configurations of its all legs, depending on the changes of the ground. It has a positive effect on the robot's stable position. Concluding, it should be emphasized that the presented control algorithm also works for $\Delta x_R(t), \Delta y_R(t), \Delta z_R(t) \neq 0$ and $\alpha_R(t), \beta_R(t), \gamma_R(t) \neq 0$. As a result, the considered construction can play a role of a mobile Stewart platform also on unstable ground.

5.3 Control of the robot standing on stable and vibrating ground

In this subsection, we considered the control problem of the robot, which is standing on vibrating ground. As in previous subsection, to better illustrate the process of controlling individual legs of the robot, we also assumed that $\Delta x_R(t) = \Delta y_R(t) = \Delta z_R(t) = 0$ and $\alpha_R(t) = \beta_R(t) = \gamma_R(t) = 0$ (i.e., full, both linear and angular, spatial stabilization of the robot's trunk). In turn, we have taken non-zero harmonic excitations of the ground, i.e., $\Delta x_G(t) \neq 0$, $\Delta y_G(t) \neq 0$, $\Delta z_G(t) \neq 0$, and $\alpha_G(t) \neq 0$, $\beta_G(t) \neq 0$, $\gamma_G(t) \neq 0$. Simulation results captured in regular time intervals are presented in **Figure 6**. As can be seen, as in previous cases, in each time the robot is supported by all eight legs, thanks to the appropriate changing of the configurations of its all legs, depending on the vibrating ground. As a result, the robot keeps its position and orientation in the global coordinate system, regardless of the vibrating ground, also increasing its stability. The presented control algorithm also works for $\Delta x_R(t), \Delta y_R(t), \Delta z_R(t) \neq 0$ and $\alpha_R(t), \beta_R(t), \gamma_R(t) \neq 0$. As a result, the considered construction can play a role of a mobile Stewart platform also on vibrating ground.

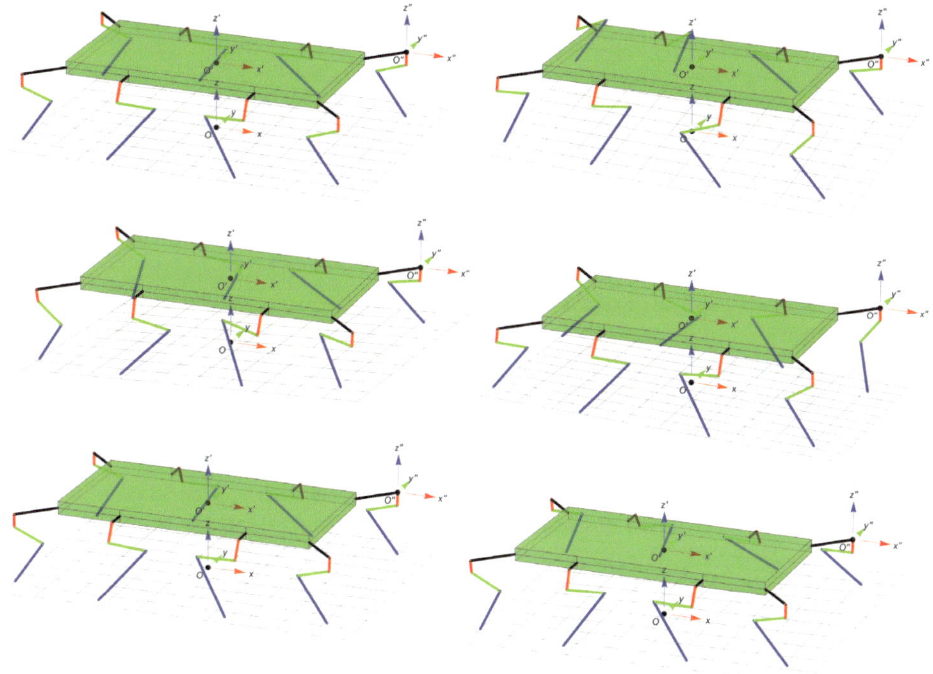

Figure 6.
Configurations of the investigated hybrid octopod robot on vibrating ground, captured in regular time intervals.

6. Application of the developed simulation model

The proposed algorithm for controlling all robot's legs during standing or walking for different cases was tested using a simulation model created in the Mathematica software. The possibilities of the Mathematica concerning 3D animation of simulated objects are available in the recent versions of this environment (in this paper we used Mathematica 11.2 version). Thanks to the developed visualization, we can observe exactly the configuration of all robot legs at any time, and this can be treated as a virtual experiment. The most recent versions of the Mathematica program also allow relatively simple operation of different microcontrollers connected to a computer such as Arduino Uno or Raspberry Pi. Moreover, simpler versions of Mathematica can also be installed in the internal memory of the mentioned Raspberry Pi microcontroller. As a result, using both the appropriate microcontroller and motor drivers, it is possible to directly apply the calculated articulated variables into joints of all robot's legs. It is schematically shown in **Figure 7**.

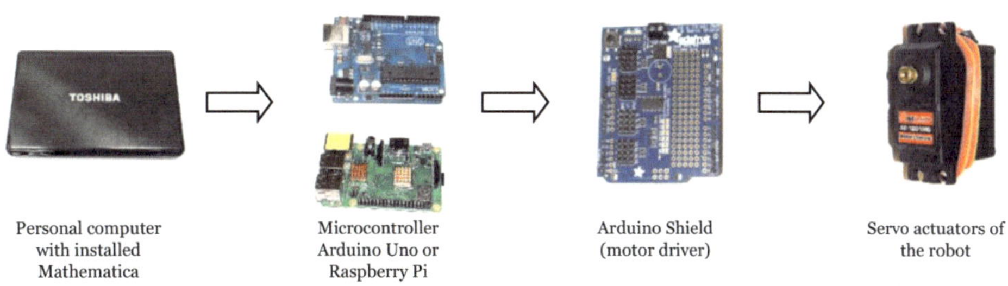

Figure 7.
Control the robots' servos of a robot prototype using a simulation model developed in Mathematica.

These tasks will be the goal of our further research related to the problem of control of the robot's legs, using real constructions made on the basis of the designs shown in **Figure 1**.

7. Conclusions

In this paper, a general model of a multi-legged robot has been developed and numerically investigated in Mathematica environment. The investigated robot was driven by a novel generator of gait proposed in our previous study. Using the developed simulation model, we performed some virtual experiments regarding locomotion process and stabilization of the position and orientation of the robot on unstable and vibrating ground. The obtained results show that the presented control possibility can be employed for full control of the robot position and orientation in space, also during locomotion of the robot. As a result, the robot can be used as a fully control walking Stewart platform. The most recent versions of Mathematica software allow to communicate with different microcontrollers, including Arduino Uno and Raspberry Pi. As a result, the developed simulation model can be relatively simply and successfully adopted to control the constructed prototypes of different kinds of multi-legged machines.

Acknowledgements

The work has been partially supported by the National Science Centre of Poland under the grant OPUS 9 no. 2015/17/B/ST8/01700 for years 2016–2018.

Author details

Dariusz Grzelczyk* and Jan Awrejcewicz
Department of Automation, Biomechanics and Mechatronics, Lodz University of Technology, Lodz, Poland

*Address all correspondence to: dariusz.grzelczyk@p.lodz.pl

References

[1] Agheli M, Qu L, Nestinger SS. SHeRo: Scalable hexapod robot for maintenance, repair, and operations. Robotics and Computer-Integrated Manufacturing. 2014;**30**(5):478-488. DOI: 10.1016/j.rcim.2014.03.008

[2] Tedeschi F, Carbone G. Design of a novel leg-wheel hexapod walking robot. Robotics. 2017;**6,40**:18. DOI: 10.3390/robotics6040040

[3] Lagaza K, Pandey A. A literature review on motion planning of hexapod machines using different soft computing methods. Global journal of engineering, science and social. Science Studies. 2018;**3**(1):1-10

[4] Grzelczyk D, Stańczyk B, Awrejcewicz J. On the hexapod leg control with nonlinear stick-slip vibrations. Applied Mechanics and Materials. 2015;**801**:12-24. DOI: 10.4028/www.scientific.net/AMM.801.12

[5] Grzelczyk D, Stańczyk B, Awrejcewicz J. Power consumption analysis of different hexapod robot gaits. In: Awrejcewicz J, Kaźmierczak M, Mrozowski J, Olejnik P, editors. Dynamical Systems - Mechatronics and Life Sciences. Lodz: TU of Lodz; 2015. pp. 197-206 (ISBN: 978-83-7283-707-3)

[6] Grzelczyk D, Stańczyk B, Awrejcewicz J. Estimation on the contact forces between the hexapod legs and the ground during walking in the tripod gait. Vibrations in Physical Systems. 2016;**27**:107-114

[7] Grzelczyk D, Stańczyk B, Szymanowska O, Awrejcewicz J. Simulation of the octopod robot controlled by different central patterns generators. In: Awrejcewicz J, Kaźmierczak M, Mrozowski J, Olejnik P, editors. Engineering Dynamics and Life Sciences. Lodz: DAB&M of TUL Press; 2017. pp. 229-238. ISBN: 978-83-935312-4-0

[8] Grzelczyk D, Szymanowska O, Awrejcewicz J. Kinematic and dynamic simulation of an octopod robot controlled by different central pattern generators. Proceedings of the Institution of Mechanical Engineers, Part I: Journal of Systems and Control Engineering. 2019;**233**(4):400-417. DOI: 10.1177/0959651818800187

[9] Grzelczyk D, Awrejcewicz J. Analysis of contact forces between the ground and the hexapod robot legs during tripod gait. Machine Dynamics Research. 2018;**42**(2):17-26

[10] Grzelczyk D, Awrejcewicz J. Modeling and control of an eight-legged walking robot driven by different gait generators. International Journal of Structural Stability and Dynamics. 2019;**19**(5):1941009-1-1941009-23. DOI: 10.1142/S0219455419410098

[11] Grzelczyk D, Awrejcewicz J. Dynamics, stability analysis and control of a mammal-like octopod robot driven by different central pattern generators. Journal of Computational Applied Mechanics. 2019;**50**(1):76-89. DOI: 10.22059/JCAMECH.2019.278583.375

[12] Rong X, Li Y, Ruan J, Li B. Design and simulation for a hydraulic actuated quadruped robot. Journal of Mechanical Science and Technology. 2012;**26**(4):1171-1177. DOI: 10.1007/s12206-012-0219-8

[13] Burkus E, Odry P. Mechanical and walking optimization of a hexapod robot using PSO. In: Proceedings of the IEEE 9th International Conference on Computational Cybernetics (ICCC '13), 8–10 July 2013, Tihany, Hungary, Art. no. 6617583. 2013:177-180

[14] Linnemann R, Klaassen B, Kirchner F. Walking robot scorpion – Experiences with a full parametric model. In: Proceedings of the European Simulation and Modelling Conference (ESM '01), 6–9 June 2001, Prague, Czech Republic. 2001. p. 7

[15] Guan X, Zheng H, Zhang X. Biologically inspired quadruped robot biosbot: Modeling, simulation and experiment. In: Proceedings of the 2nd International Conference on Autonomous Robots and Agents, 13–15 December 2004, Palmerston North, New Zealand. pp. 261-266, 2004

[16] Chen W, Ren G, Zhang J, Wang J. Smooth transition between different gaits of a hexapod robot via a central pattern generators algorithm. Journal of Intelligent and Robotic Systems. 2012;**67** (3–4):255-270. DOI: 10.1007/s10846-012-9661-1

[17] Koo IM, Tran DT, Kang TH, Vo GL, Song YK, Lee CM, et al. Control of a quadruped walking robot based on biologically inspired approach. In: Proceedings of the 2007 IEEE/RSJ International Conference on Intelligent Robots and Systems (IROS '07), October 29–November 2, 2007; San Diego, CA, USA. 2007. pp. 2969-2974

[18] Koo IM, Kang TH, Vo GL, Trong TD, Song YK, Choi HR. Biologically inspired control of quadruped walking robot. International Journal of Control, Automation and Systems. 2009;**7**(4):577-584. DOI: 10.1007/s12555-009-0409-z

[19] Mahapatra A, Roy SS. Computer aided dynamic simulation of six-legged robot. International Journal of Recent Trends in Engineering. 2009;**2**(2): 146-151

[20] Cekus D, Kubik L, Piotrowski Ł, Skrobek D. The application of SolidWorks to conduct advanced simulations of virtual prototypes. Engineering Modelling. 2017;**63**:32-40 (in Polish)

[21] Ptak P, Pierzgalski M, Cekus D, Sokół K. Modeling and stress analysis of a frame with a suspension of a Mars rover. Procedia Engineering. 2017;**177**: 175-181. DOI: 10.1016/j.proeng.2017.02.215

[22] Lewis MA, Bekey GA. Gait adaptation in a quadruped robot. Autonomous Robots. 2002;12(3):301-312. DOI: 10.1023/A:1015221832567

[23] Pongas D, Mistry M, Schaal S. A robust quadruped walking gait for traversing rough terrain. In: Proceedings of the IEEE International Conference on Robotics and Automation, 10–14 April 2007, Rome, Italy. 2007. pp. 1474-1479

[24] Bo J, Cheng C, Wei L, Xiangyun L. Design and configuration of a hexapod walking robot. In: Proceedings of the Third International Conference on Measuring Technology and Mechatronics Automation, 6–7 January 2011, Shanghai, China. 2011. pp. 863-866

[25] Lee YH, Tran DT, Hyun J-h, Phan LT, Koo IM, Yang SU, et al. A gait transition algorithm based on hybrid walking gait for a quadruped walking robot. Intelligent Service Robotics. 2015;**8**(4):185-200. DOI: 10.1007/s11370-015-0173-2

[26] Chen G, Jin B, Chen Y. Tripod gait-based turning gait of a six-legged walking robot. Journal of Mechanical Science and Technology. 2017;**31**(3): 1401-1411. DOI: 10.1007/s12206-017-0241-y

[27] Nandhini M, Krithika V, Chittal K. Design of four pedal quadruped robot. In: Proceedings of the IEEE International Confs. 2017. pp. 2548-2552

Chapter 7

On Dynamics and Invariant Sets in Predator-Prey Maps

Blai Vidiella, J. Tomás Lázaro, Lluís Alsedà and Josep Sardanyés

Abstract

A multitude of physical, chemical, or biological systems evolving in discrete time can be modelled and studied using difference equations (or iterative maps). Here we discuss local and global dynamics for a predator-prey two-dimensional map. The system displays an enormous richness of dynamics including extinctions, co-extinctions, and both ordered and chaotic coexistence. Interestingly, for some regions we have found the so-called hyperchaos, here given by two positive Lyapunov exponents. An important feature of biological dynamical systems, especially in discrete time, is to know where the dynamics lives and asymptotically remains within the phase space, that is, which is the invariant set and how it evolves under parameter changes. We found that the invariant set for the predator-prey map is very sensitive to parameters, involving the presence of escaping regions for which the orbits go out of the domain of the system (the species overcome the carrying capacity) and then go to extinction in a very fast manner. This theoretical finding suggests a potential dynamical fragility by which unexpected and sharp extinctions may take place.

Keywords: bifurcations, chaos, invariant sets, maps, nonlinearity, ecology

1. Introduction

Natural and artificial complex systems can evolve in discrete time, often resulting in extremely complex dynamics such as chaos. A well-known example of such a complexity is found in ecology, where discrete-time dynamics given by a yearly climatic forcing can make the population emerging a given year to be a discrete function of the population of the previous one [1]. Although early work already pointed towards complex population fluctuations as an expected outcome of the nonlinear nature of species interactions [2], the first evidence of chaos in species dynamics was not characterised until the late 1980s and 1990s [3, 4]. Since pioneering works on one-dimensional maps [5, 6], the field of dynamical complexity in ecology experienced a rapid development [5–7], with several key investigations offering a compelling evidence of chaotic dynamics in insect species in nature [1, 3, 4].

Discrete-time models have played a key role in the understanding of complex ecosystems, especially for univoltine species (i.e. species undergoing one generation per year) [5, 6]. Many insects inhabiting temperate and boreal climatic zones

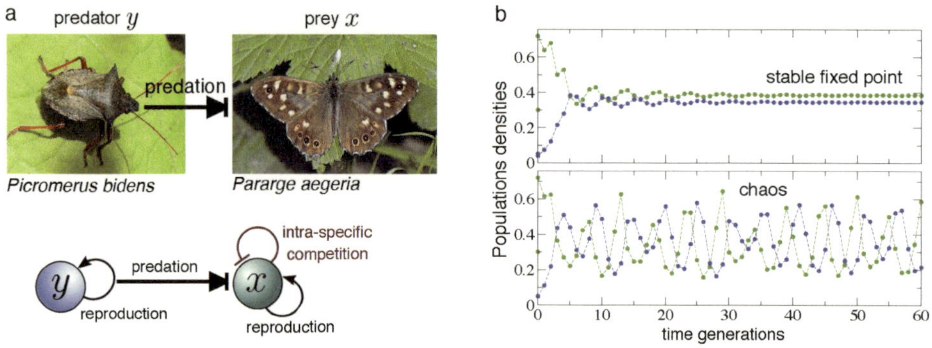

Figure 1.
Two-species predator-prey dynamics can be studied with difference equations or maps when species generations are discrete (univoltine). (a) Here we display two insect species with univoltine generations at the North Hemisphere. The Heteroptera Picromerus bidens predates the butterfly Pararge aegeria by consuming the eggs (photos obtained from the Wikipedia). A simple model for this type of system is given by the map (1). (b) Some typical dynamics arising in discrete-time ecological systems for preys (green dots) and predators (blue dots): (upper panel) period-one fixed point and (lower panel) chaos.

behave as univoltine species, for example, *Lepidoptera* [8], *Coleoptera* [9], or *Heteroptera* [10] species, among others. For *Lepidoptera*, the populations of the butterfly *Pararge aegeria* are univoltine in its most northern range (e.g. northern Scandinavia). Adult butterflies emerge in late spring, mate, and die shortly after laying the eggs. Then, their offspring grow until pupation, entering diapause before winter. New adults emerge the following year, thus resulting in a single generation of butterflies per year [11].

Some predators feed on these univoltine insects. For example, *Picromerus bidens* (*Heteroptera*) predates on *Pararge aegeria* by consuming their eggs. Thus, both prey and predator display coupled yearly cycles (**Figure 1(a)**). This type of systems has been modelled using two-dimensional discrete-time models, such as the one we are introducing in this chapter, given by the map (1) (see Ref. [12] for more details on this model). As mentioned, the dynamical richness of discrete ecological models was early recognised [5, 6] and special attention has been paid to small food chains incorporating two species in discrete systems [12]. These systems, similarly to single-species maps, display static equilibria, periodic population oscillations, as well as chaotic dynamics (see, e.g. **Figure 1(b)**).

A crucial point that we want to address in this chapter is the proper characterisation of the invariant set in which the dynamics lives. This is of paramount importance for discrete-time systems since the iterates can undergo big jumps within the phase space and extinctions can occur in a very catastrophic manner if some iterate visits the so-called escaping regions. That is, catastrophic extinctions not caused by bifurcations but from topological features of the invariant sets may occur. Together with the characterisation of the invariant set, we provide a dynamical analysis of fixed points, local and global stability, as well as a numerical investigation of chaos.

2. Predator-prey map

We consider a food chain of two interacting species with predator-prey dynamics, each with nonoverlapping generations (see **Figure 1(a)**). The preys x grow logistically without the presence of predators population y, following the logistic map [6]. The proposed model to study such ecosystem can be described by the following system of nonlinear difference equations [12]:

$$\begin{pmatrix} x_{n+1} \\ y_{n+1} \end{pmatrix} = T\begin{pmatrix} x_n \\ y_n \end{pmatrix} \quad \text{where} \quad T\begin{pmatrix} x \\ y \end{pmatrix} = T_{\mu,\beta}\begin{pmatrix} x \\ y \end{pmatrix} = \begin{pmatrix} \mu x(1-x-y) \\ \beta xy \end{pmatrix} \quad (1)$$

is defined on the phase space given by the simplex:

$$S = \{(x,y) : x,y \geq 0 \text{ and } x+y \leq 1\}.$$

We will focus our analysis on the parameter regions, $\mu \in (0,4]$ and $\beta \in (0,5]$, which contain relevant biological dynamics. State variables $(x,y) \in [0,1]^2$ denote population densities with respect to a normalised carrying capacity for preys ($K = 1$). Observe that, in fact, if we do not normalise the carrying capacity, the term $1-x-y$ in $T_{\mu,\beta}$ should read $1-x/K-y$. As mentioned, preys grow logistically with an intrinsic reproduction rate $\mu > 0$ without predators. Finally, preys' reproduction is decreased by the action of predators, which increase their population numbers at a rate $\beta > 0$ due to consumption of preys.

3. Fixed points and local stability

The next lemma provides the three fixed points of the dynamical system defined by the map (1) for $(\mu,\beta) \in (0,4] \times (0,5]$ and the parameter regions for which they belong to the simplex S.

Lemma 1.1. The dynamical system (1) on the simplex S has the following three fixed points (see **Figure 2** (left)):

- $P_1^* = (0,0)$ which belongs to the simplex S for every (μ,β).
- $P_2^* = \left(1-\frac{1}{\mu},0\right)$ which belongs to the simplex S for every $(\mu,\beta) \in [1,4] \times (0,5]$.
- $P_3^* = \left(\frac{1}{\beta},1-\frac{1}{\mu}-\frac{1}{\beta}\right)$ which belongs to the simplex S for every

$$(\mu,\beta) \in \left[\frac{5}{4},4\right] \times \left[\frac{\mu}{\mu-1},5\right].$$

The fixed point P_1^* corresponds to co-extinctions, P_2^* to predator extinction and prey survival, and P_3^* to the coexistence of both populations.

Proof: It is a routine to check that P_1^*, P_2^* and P_3^* are the unique possible fixed points of model (1). Thus, the first and the second statements of the lemma are evident.

We need to prove that P_3^* belongs to the simplex S if and only if $(\mu,\beta) \in \left[\frac{5}{4},4\right] \times \left[\frac{\mu}{\mu-1},5\right]$.

Observe that the inequalities $\mu > 0$ and $\beta > 0$ directly give $\frac{1}{\beta} > 0$, $1-\frac{1}{\mu}-\frac{1}{\beta} < 1$, and $1-\frac{1}{\mu} = \frac{1}{\beta} + \left(1-\frac{1}{\mu}-\frac{1}{\beta}\right) < 1$. So, the statement $P_3^* \in S$ is equivalent to $\frac{1}{\beta} < 1$ and

$$0 \leq 1-\frac{1}{\mu}-\frac{1}{\beta} \Leftrightarrow \begin{cases} \mu > 1, \text{and} \\ \frac{1}{\beta} \leq 1-\frac{1}{\mu} = \frac{\mu-1}{\mu} \end{cases} \Leftrightarrow \begin{cases} \mu > 1, \ \text{and} \\ \beta \geq \frac{\mu}{\mu-1}. \end{cases}$$

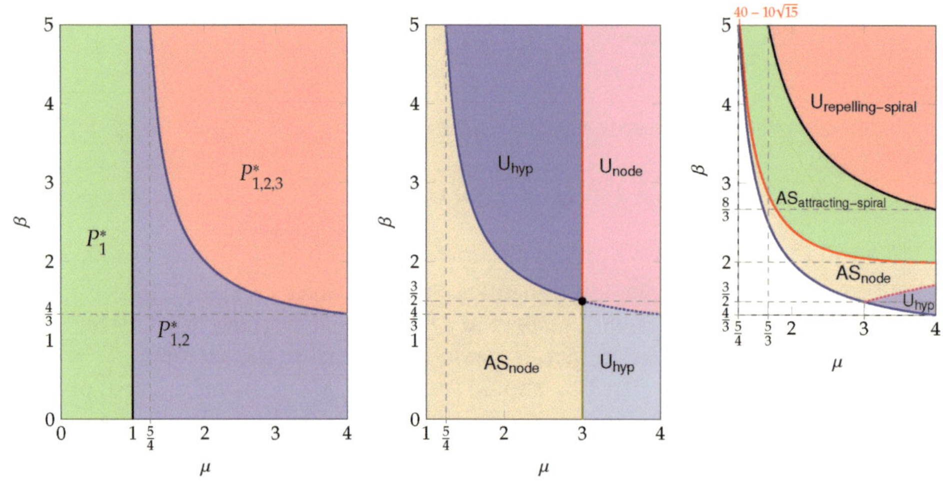

Figure 2.
The left picture shows the regions of existence of the fixed points $P^*_{1,2,3}$. The centre (respectively right) picture specifies the regions of the parameter space (of course in its parametric domain of definition) where the fixed point P^*_2 (respectively P^*_3) has different local dynamics, together with the type of local dynamics displayed in each of the regions. The analogous picture for the point P^*_1 has been omitted for simplicity. The codification for the stability zones follows the next rules: Capital letters indicate stability—U indicates unstable, while AS indicates asymptotic stability. The subscripts show the type of stability: hyp = hyperbolic, node, attracting-spiral, and repelling-spiral.

Clearly, the last two conditions give $\beta \geq \frac{\mu}{\mu-1} > 1$ which is equivalent to $\frac{1}{\beta} < 1$. On the other hand, $\frac{\mu}{\mu-1} \leq \beta \leq 5$ is equivalent to $\mu \geq \frac{5}{4}$.

In the next three lemmas, the different regions of local stability of these fixed points are studied. This study, standard in dynamical systems theory, is based on the computation of the eigenvalues of the Jacobian matrix at each fixed point and on the determination of the regions where their moduli are smaller or larger than 1. To ease the reading, the proofs have been deferred to the end of the section.

Lemma 1.2 (Stability of the point P_1^*) The fixed point P_1^* is locally asymptotically stable (of attractor node type) if $\mu \in (0,1)$, with eigenvalues $\lambda_1 = \mu < 1$, $\lambda_2 = 0$, and unstable (of hyperbolic type) if $\mu \in (1,4]$. In that case its eigenvalues are $\lambda_1 = \mu > 1$ and $\lambda_2 = 0$.

Observe that in both cases, there is an eigendirection, corresponding to the y-axis, which is strongly attracting. As it often happens in many biological systems, its change of stability coincides with the "birth" of the fixed point P_2^*.

Lemma 1.3 (Stability of the point P_2^*). Let us consider in the parameter region $(\mu, \beta) \in [1,4] \times (0,5]$, the domain of existence of the fixed point $P_2^* \in S$, the curve

$$\beta = \frac{\mu}{\mu - 1} \tag{2}$$

(defined and contained in the domain for $\mu \geq \frac{5}{4}$), and the vertical line $\mu = 3$. The curve and the line divide this domain into four regions (as shown in **Figure 2** (centre)). Then, the local stability of system (1) in a neighbourhood of the fixed point P_2^* is as follows: In the bottom-left region (brown), it is locally asymptotically stable (attractor of node type). In the top-right region (magenta), it is unstable (repelling of node type). In the bottom-right and top-left regions (in light blue colour), P_2^* is also unstable, but of hyperbolic type. In the bottom part, the eigenvalues satisfy $|\lambda_1| > 1$ and $|\lambda_2| < 1$, while in the top part, these inequalities are reversed, $|\lambda_1| < 1$ and $|\lambda_2| > 1$. As usual, the curves and lines defining the border between these regions are characterised by a pass-through modulus 1 of some of the eigenvalues. Indeed, on the curve (2) (in blue colour, solid and dashed), one has $\lambda_2 = 1$, and on the vertical line $\mu = 3$ (in red and green colours), one gets $\lambda_1 = -1$. On the **black** point at the intersection of both curves, which has coordinates $(\mu, \beta) = (3, 1.5)$, the eigenvalues are $\lambda_1 = -1$ and $\lambda_2 = 1$.

And last but not least, the following lemma establishes the different regions of stability for the point P_3^*, the coexistence equilibrium.

Lemma 1.4 (Stability of the point P_3^*) Let us consider in the parameter region $(\mu, \beta) \in \left[\frac{5}{4}, 4\right] \times \left[\frac{\mu}{\mu-1}, 5\right]$, the domain of existence of the fixed point $P_3^* \in S$, the above curve (2), and the following three curves:

$$\beta = 2\frac{\mu}{\mu - 1} \quad \text{(black)}, \tag{3}$$

$$\beta = \frac{\mu}{2\left(\sqrt{\mu} - 1\right)} \quad \text{(red)}, \tag{4}$$

$$\beta = 3\frac{\mu}{\mu + 3} \quad \text{(dashed magenta)}. \tag{5}$$

These curves divide this domain into four regions (see **Figure 2** (right)):

1. The region at the top, coloured in pink and delimited by the curve (3), where the point P_3^* is unstable of repeller spiral type (its Jacobian matrix has complex eigenvalues with $|\lambda_{1,2}| > 1$).

2. The green-coloured zone, delimited by the curves (3) and (4), where P_3^* is asymptotically stable of attracting spiral type with complex eigenvalues satisfying $|\lambda_{1,2}| < 1$.

3. The region in brown colour, delimited by the curves (2), (4), and (5). Here the Jacobian matrix of P_3^* has real eigenvalues with $|\lambda_{1,2}| < 1$, and P_3^* is locally asymptotically stable of node type.

4. The bottom region, in light blue, where $|\lambda_1| < 1$ and $\lambda_2 < -1$. Therefore, P_3^* is unstable of hyperbolic type.

We present now the proofs of Lemma 1.3 and Lemma 1.4. The one of Lemma 1.2 has been omitted since it consists on straightforward computations.

Proof of Lemma 1.3: The Jacobian matrix of T at the point P_2^* is

$$DT\left(P_2^*\right) = \begin{pmatrix} 2-\mu & 1-\mu \\ 0 & \beta\left(1 - \dfrac{1}{\mu}\right) \end{pmatrix},$$

being triangular, so its eigenvalues are $\lambda_1 = 2 - \mu$ and $\lambda_2 = \beta\left(1 - \frac{1}{\mu}\right)$. They are both real and, since $\mu \in (1, 4]$, λ_2 is positive, and concerning λ_1, one has $|\lambda_1| < 1$ when $\mu \in (1, 3)$, $|\lambda_1| = 1$ when $\mu = 3$, and $|\lambda_1| > 1$ when $\mu \in (3, 4]$. To determine more precisely the local stability of P_2^*, we study the modulus of λ_2 on each of these intervals.

Case $\mu \in (1, 3)$. As we already said, in this case we have $|\lambda_1| < 1$ and $\lambda_2 > 0$. The curve $\lambda_2 = 1$ is the curve (2) (in solid blue colour in **Figure 2** (centre)). This curve intersects the line $\mu = 3$ at $\beta = 3/2$ and the line $\beta = 5$ at $\mu = 5/4$. On this curve the linearised system is stable but nothing can be said, a priori, about the nonlinear system. For the parameters β and μ for which $\beta > \frac{\mu}{\mu-1}$, we have $\lambda_2 > 1$ and, hence, P_2^* is unstable of hyperbolic type. In a similar way, for those parameters verifying $\beta < \frac{\mu}{\mu-1}$, we get that both eigenvalues $\lambda_{1,2}$ have modulus strictly smaller than 1. Hence, P_2^* is asymptotically stable of node type.

Case $\mu = 3$. Now the eigenvalues are $\lambda_1 = -1$ and $\lambda_2 = \frac{2\beta}{3}$. When $\beta = \frac{3}{2}$, $\lambda_1 = -1$, and $\lambda_2 = 1$, so P_2^* is stable for the linearised system. Notice that $(\mu, \beta) = (3, 3/2)$ is exactly the intersection point of the curve (2) with the line $\mu = 3$. If $\beta > \frac{3}{2}$, then $\lambda_1 = -1$ and $\lambda_2 > 1$, so P_2^* is unstable. Finally, if $\beta < \frac{3}{2}$, then $\lambda_1 = -1$ and $\lambda_2 < 1$, and therefore P_2^* is stable for the linearised system.

Case $\mu \in (3, 4]$. Since $|\lambda_1| > 1$, the point P_2^* is always unstable. Moreover, as in the case $\mu \in (1, 3)$, the modulus of λ_1 depends on the position of μ and β with respect to the curve (2) ($\lambda_2 = 1$). Consequently, if $\beta = \frac{\mu}{\mu-1}$, then $\lambda_2 = 1$ and P_2^* is unstable. If $\beta > \frac{\mu}{\mu-1}$, then $\lambda_2 > 1$ and P_2^* is unstable (of node type). Finally, if $\beta < \frac{\mu}{\mu-1}$, then $|\lambda_2| < 1$ and P_2^* is an (unstable) hyperbolic point.

Proof of Lemma 1.4: The Jacobian matrix of T at the point P_3^* is

$$DT(P_3^*) = \begin{pmatrix} 1 - \dfrac{\mu}{\beta} & -\dfrac{\mu}{\beta} \\ \beta\left(1 - \dfrac{1}{\mu} - \dfrac{1}{\beta}\right) & 1 \end{pmatrix}.$$

Then, the trace, the determinant of $DT(P_3^*)$ and the discriminant of the characteristic polynomial of this matrix are

$$\tau = \mathrm{tr}DT(P_3^*) = 2 - \frac{\mu}{\beta}, \qquad D = \det\, DT(P_3^*) = \mu\left(1 - \frac{2}{\beta}\right), \text{ and} \tag{6}$$

$$\Delta = \tau^2 - 4D = \left(2 - \frac{\mu}{\beta}\right)^2 - 4\mu\left(1 - \frac{2}{\beta}\right) = \left(\frac{\mu}{\beta} + 2\right)^2 - 4\mu. \tag{7}$$

The eigenvalues of $DT(P_3^*)$ are given by

$$\lambda_{1,2} = \frac{\tau \pm \sqrt{\Delta}}{2}. \tag{8}$$

The curve determining whether the eigenvalues are real or complex is $\Delta = 0$, that is,

$$\Delta = 0 \quad \Leftrightarrow \quad \left(\frac{\mu}{\beta} + 2\right)^2 = 4\mu \quad \Leftrightarrow \quad \frac{\mu}{\beta} = 2(\sqrt{\mu} - 1) \quad \Leftrightarrow \quad \beta = \frac{\mu}{2(\sqrt{\mu} - 1)},$$

which corresponds to the red curve (4).

Observe that in the region above the red curve (4), $\Delta < 0$. So, the stability of P_3^* in this region is determined by the modulus of

$$\lambda_{1,2} = \frac{\tau \pm \mathrm{i}\sqrt{-\Delta}}{2} = \frac{\left(2 - \frac{\mu}{\beta}\right) \pm \mathrm{i}\sqrt{4\mu - \left(\frac{\mu}{\beta} + 2\right)^2}}{2}.$$

Precisely, we are interested on determining when $|\lambda_{1,2}| = 1$ or, equivalently, when $|\lambda_{1,2}|^2 = 1$. We have

$$|\lambda_{1,2}|^2 = \frac{\left(2-\frac{\mu}{\beta}\right)^2 + 4\mu - \left(2+\frac{\mu}{\beta}\right)^2}{4} = \frac{4\mu - 8\frac{\mu}{\beta}}{4} = \mu\left(1-\frac{2}{\beta}\right).$$

Therefore, $1 = |\lambda_{1,2}|^2 = \mu\left(1-\frac{2}{\beta}\right)$ is equivalent to $\beta = \frac{2\mu}{\mu-1}$, which is the black curve (3). This implies that, in the pink-coloured region above the black curve (3), displayed in **Figure 2** (right), the point P_3^* has complex eigenvalues with modulus greater than 1, and, consequently, it is unstable of repelling spiral type. Analogously, the green region corresponds to complex eigenvalues $\lambda_{1,2}$, with (both) moduli smaller than 1. Here, P_3^* is asymptotically stable of attracting spiral type.

In the region below the red curve (4), where $\Delta > 0$, both eigenvalues are real. They can be rewritten as

$$\lambda_{1,2} = \left(1-\frac{\mu}{2\beta}\right) \pm \sqrt{\left(\frac{\mu}{2\beta}+1\right)^2 - \mu},$$

being λ_1 (respectively λ_2) the eigenvalue corresponding to the + (respectively −) sign.

First we will show that $|\lambda_1(\mu, \beta)| < 1$ in the region delimited by the curves (4) and (2) (including the graph of the curve (4) and excluding the graph of the curve (2)). Observe that, since $\frac{\mu}{\mu-1} \leq \beta$ and $\mu \leq 4$, we have

$$\frac{\mu}{2\beta} - 2 \leq \frac{\mu-1}{2} - 2 < 0 \leq \sqrt{\left(\frac{\mu}{2\beta}+1\right)^2 - \mu} \Leftrightarrow -1 < \left(1-\frac{\mu}{2\beta}\right) + \sqrt{\left(\frac{\mu}{2\beta}+1\right)^2 - \mu}$$
$$= \lambda_1.$$

Furthermore,

$$1 > \lambda_1 = \left(1-\frac{\mu}{2\beta}\right) + \sqrt{\left(\frac{\mu}{2\beta}+1\right)^2 - \mu} \Leftrightarrow \sqrt{\left(\frac{\mu}{2\beta}+1\right)^2 - \mu} < \frac{\mu}{2\beta} \Leftrightarrow \left(\frac{\mu}{2\beta}+1\right)^2$$
$$-\mu < \left(\frac{\mu}{2\beta}\right)^2 \Leftrightarrow 1 + \frac{\mu}{\beta} < \mu \Leftrightarrow \beta > \frac{\mu}{\mu-1},$$

This proves that, indeed, $|\lambda_1(\mu, \beta)| < 1$ in the region delimited by the curves (4) and (2), excluding the graph of the curve (2).

Now we study $|\lambda_2|$. Observe that, clearly,

$$-\sqrt{\left(\frac{\mu}{2\beta}+1\right)^2 - \mu} \leq 0 < \frac{\mu}{2\beta} \quad \Leftrightarrow \quad \lambda_2 = \left(1-\frac{\mu}{2\beta}\right) - \sqrt{\left(\frac{\mu}{2\beta}+1\right)^2 - \mu} < 1.$$

Next, by using again that $\frac{\mu}{2\beta} - 2 < 0$, we have

$$-1 = \lambda_2 = \left(1-\frac{\mu}{2\beta}\right) - \sqrt{\left(\frac{\mu}{2\beta}+1\right)^2 - \mu} \Leftrightarrow \frac{\mu}{2\beta} - 2$$
$$= -\sqrt{\left(\frac{\mu}{2\beta}+1\right)^2 - \mu} \Leftrightarrow \sqrt{\left(\frac{\mu}{2\beta}+1\right)^2 - \mu} = 2 - \frac{\mu}{2\beta} \Leftrightarrow \left(\frac{\mu}{2\beta}+1\right)^2 - \mu$$
$$= \left(2-\frac{\mu}{2\beta}\right)^2 \Leftrightarrow 3\frac{\mu}{\beta} = \mu + 3 \Leftrightarrow \beta = 3\frac{\mu}{\mu+3}.$$

The last equality is curve (5) and, as shown in **Figure 2** (right), it intersects the curve (2) at the point $(\mu, \beta) = (3, 3/2)$, it is strictly increasing in the interval $\mu \in [3, 4]$, and intersects the line $\mu = 4$ at $\beta = 12/7 < 2$. By using the above chain of equivalent equalities, it is easy to check that $\lambda_2 > -1$ if and only if $\beta > 3\frac{\mu}{\mu+3}$. Thus, the assertions (3) and (4) of the lemma follow straightforwardly.

4. Invariant set: where dynamics live and remain

A first natural question is whether and when S is the domain of the dynamical system associated with model (1). This amounts asking whether and when S is T-invariant (i.e. $T(\mathrm{S}) \subset \mathrm{S}$). The complete answer to this question is given by the following proposition and corollary.

In this section, at some point we will consider $\mu(\beta)$ as a function of β. So, for consistency, instead of using the simple notation T for the map from model (1), we will use the notation $T_{\mu,\beta}$ which emphasises the explicit dependence of T on the two parameters μ and β.

Proposition 1.5 $T_{\mu,\beta}(\mathrm{S}) = \left\{(x,y) \in \mathbb{R}^+ \times \mathbb{R}^+ : \frac{x}{\mu} + \frac{y}{\beta} \leq \frac{1}{4}\right\}$.

Remark 1.6 Indeed, we can say more: any point $(u,v) \in \mathbb{R}^+ \times \mathbb{R}^+$ such that

$$\frac{u}{\mu} + \frac{v}{\beta} < \frac{1}{4}$$

admits, exactly, two $T_{\mu,\beta}$ preimages, and they belong to S. Moreover, if $(u,v) \in \mathbb{R}^+ \times \mathbb{R}^+$ is such that $\frac{u}{\mu} + \frac{v}{\beta} = \frac{1}{4}$, then $\left(\frac{1}{2}, \frac{2v}{\beta}\right) \in \mathrm{S}$ is the only $T_{\mu,\beta}$ preimage of (u,v).

The line $\frac{x}{\mu} + \frac{y}{\beta} = \frac{1}{4}$ joins the point $\left(\frac{\mu}{4}, 0\right)$ with $\left(0, \frac{\beta}{4}\right)$. So, when $\beta \leq 4$, it is below the line $x + y = 1$ and when $\beta > 4$ it has points outside S. Consequently, from Proposition 1.5 we get

Corollary 1.7 *The simplex* S *is* $T_{\mu,\beta}$*-invariant if and only if* $\beta \leq 4$.

Remark 1.8 In fact, it can be easily shown that $\beta \leq 4$ implies $T_{\mu,\beta}(\mathrm{S}) \not\subseteq \mathrm{S}$ except when $\mu = \beta = 4$.

Proof of Proposition 1.5: We start by proving that

$$T_{\mu,\beta}(\mathrm{S}) \subset \left\{(x,y) \in \mathbb{R}^+ \times \mathbb{R}^+ : \frac{x}{\mu} + \frac{y}{\beta} \leq \frac{1}{4}\right\}.$$

Let $(x,y) \in \mathrm{S}$. We have $T(x,y) = (\mu x(1-x-y), \beta xy)$ and, $\mu x(1-x-y), \beta xy \geq 0$ because $\mu, \beta > 0$ and, since $(x,y) \in \mathrm{S}$, $x, y \geq 0$ and $x + y \leq 1$. So, we have proved that $T(x,y) \in \mathbb{R}^+ \times \mathbb{R}^+$. To end the proof of the above inclusion, we have to show that $\frac{\mu x(1-x-y)}{\mu} + \frac{\beta xy}{\beta} \leq \frac{1}{4}$. We have

$$\frac{\mu x(1-x-y)}{\mu} + \frac{\beta xy}{\beta} = x(1-x-y) + xy = x(1-x) \leq \frac{1}{4}.$$

Next we will show that for every $(u,v) \in \mathbb{R}^+ \times \mathbb{R}^+$ such that $\frac{u}{\mu} + \frac{v}{\beta} \leq \frac{1}{4}$, there exists $(x,y) \in \mathrm{S}$ such that $T(x,y) = (\mu x(1-x-y), \beta xy) = (u,v)$ (i.e. $u = \mu x(1-x-y)$ and $v = \beta xy$).

If $v = 0$, it is enough to take $y = 0$ and x such that $\mu x(1-x) = u$. Observe that such point x exists because, in this case,

$$0 \le u = \mu \left[\frac{u}{\mu} + \frac{v}{\beta}\right] \le \frac{\mu}{4}.$$

Next we suppose that $v > 0$. The fact that $u \in \mathbb{R}^+$ together with $\frac{u}{\mu} + \frac{v}{\beta} \le \frac{1}{4}$ implies that $0 < v \le \frac{\beta}{4}$. So, there exist two points $0 < y^- \le \frac{1}{2} \le y^+ < 1$ such that

$$\beta y^-(1-y^-) = \beta y^+(1-y^+) = v.$$

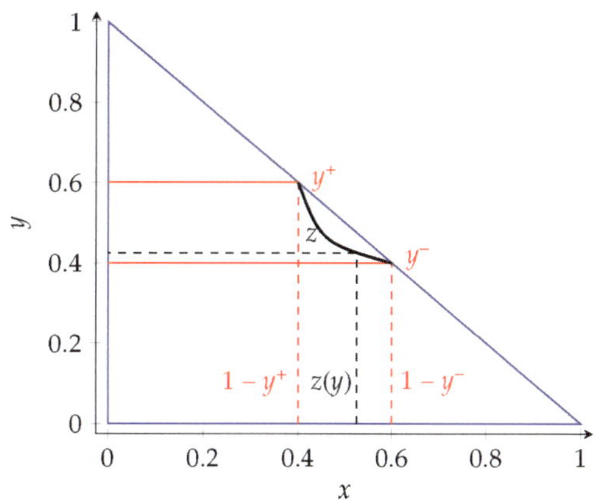

Since $\beta > 0$ and $0 < y^- \le y^+ < 1$, the function $z(y) = \frac{v}{\beta y}$ from the interval $[y^-, y^+]$ to $[1-y^+, 1-y^-]$ is a decreasing homeomorphism (observe that we have $z(y)$; $z(y^\pm) = 1 - y^\pm$). Moreover, since $y^- \le \frac{1}{2} \le y^+$, we obtain $1 - y^+ \le \frac{1}{2} \le 1 - y^-$ (see plot above). Consequently,

$$\{\beta x(1-x) : x \in [1-y^+, 1-y^-]\} = \left[v, \frac{\beta}{4}\right].$$

Hence, there exists a point $x = z(y) \in [1-y^+, 1-y^-]$ (of course with $y \in [y^-, y^+]$) such that $\beta x(1-x) = \frac{\beta}{\mu}u + v$ because $v \le \frac{\beta}{\mu}u + v \le \frac{\beta}{4}$. Then, for these particular values of y and $x = z(y)$, we have $\beta yx = v$ and

$$\mu x(1-y-x) = \mu x(1-x) - \mu yx = \frac{\mu}{\beta}(\beta x(1-x) - \beta yx) = \frac{\mu}{\beta}\left(\frac{\beta}{\mu}u + v - v\right) = u.$$

Next we consider the case $\beta > 4$. We want to find an invariant subset of S or, equivalently, the domain of definition of $T_{\mu,\beta}$ as a dynamical system.

We define the *one-step escaping set* $\varepsilon_{\mu,\beta}$ as the set of points $z \in \mathrm{S}$ such that $T_{\mu,\beta}(z) \notin \mathrm{S}$ (see **Figure** 7 for an example). Obviously, $\varepsilon_{\mu,\beta} \subset \mathrm{S}$ by definition.

The next proposition gives an estimate of the domain of definition of $T_{\mu,\beta}$ as a dynamical system (i.e. a $T_{\mu,\beta}$-invariant subdomain of S) when $\beta > 4$ and μ is small enough.

Proposition 1.9 For every $\beta > 4$, there exists a unique value $\mu^* = \mu^*(\beta) \in (0, 4)$ for which the parabola $y = \frac{1-\mu^* x(1-x)}{(\beta-\mu^*)x}$ and the line $\frac{x}{\mu^*} + \frac{y}{\beta} = \frac{1}{4}$ intersect at a unique point (see **Figure 3**). Then, the set $\mathrm{S} \backslash \varepsilon_{\mu,\beta}$ is $T_{\mu,\beta}$-invariant for every $\beta > 4$ and $\mu \le \mu^*(\beta)$.

Proposition 1.9 together with Corollary 1.7 give the splitting of the parameter space according to the shape of the invariant set. **Figure 4** and its caption give a graphical description of this splitting together with an account of some dynamical aspects in the different regions (see also **Figures 5** and **6**).

It is well known that the recurrent dynamics of a dynamical system (S, T) takes place in the *non-wandering set of* T, $\Omega(T)$, and $\Omega(T) \subset \cap_{i=0}^{\infty} T^{i}(S)$ (see, for instance,

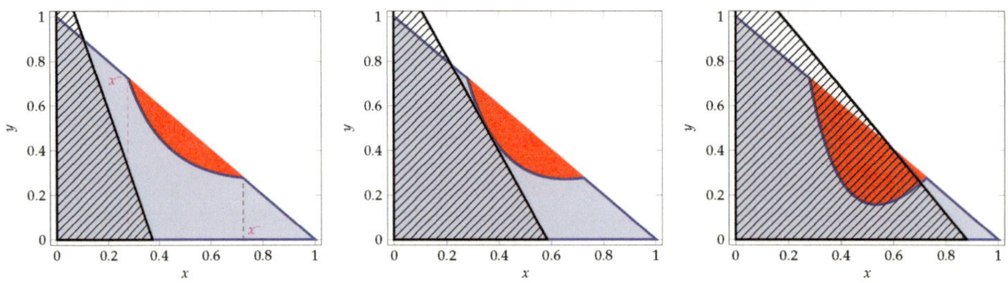

Figure 3.
Three examples of the domain $S \backslash \varepsilon_{\mu,\beta}$ *(in blue) with the one-step escaping set* $\varepsilon_{\mu,\beta}$ *plotted in red for* $\beta = 5$ *and* $\mu = 1.5$ *(left picture),* $\mu = 2.340246528387\cdots$ *(centre picture), and* $\mu = 3.525$ *(right picture). The black region shows the set* $\{(x,y) \in \mathbb{R}^{+} \times \mathbb{R}^{+} : \frac{x}{\mu} + \frac{y}{\beta} \leq \frac{1}{4}\} \supset T_{\mu,\beta}(S) \supset T_{\mu,\beta}(S \backslash \varepsilon_{\mu,\beta})$.

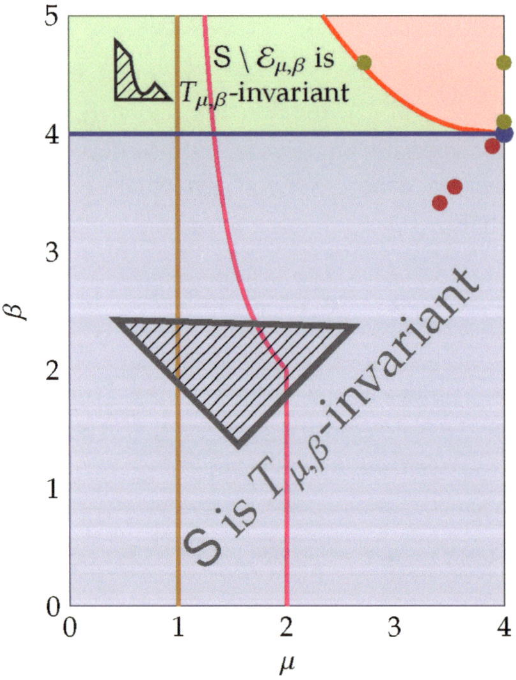

Figure 4.
The blue region is the one studied by Corollary 1.7: the set S is $T_{\mu,\beta}$*-invariant. The blue point* ($\beta = \mu = 4$) *is, according to Remark 1.8, the unique point where* $T_{\mu,\beta}(S) = S$. *The red curve is* $(\mu^{*}(\beta), \beta)$ *for* $\beta \in (4, 5]$ *(see Remark 1.10). The green region union with the red curve corresponds to Proposition 1.9:* $S \backslash \varepsilon_{\mu,\beta}$ *is* $T_{\mu,\beta}$*-invariant. The region at the left of the brown vertical line (i.e.* $\mu < 1$*) corresponds to the parameters for which there exists global convergence to the fixed point* P_1^{*} *(Theorem 1.13). The region between the line* $\mu = 1$ *and the magenta curve* $(\varphi(\beta), \beta)$ *with* $\varphi(x) := \begin{cases} 2 & \text{for } x \in [0, 2], \\ \frac{x}{x-1} & \text{for } x \in [2, 5], \end{cases}$ *corresponds to the parameters for which there exists global convergence to* P_2^{*} *(except for the escaping points and the preimages of* P_1^{*}*—Theorem 1.14). The purple dots mark the values of the parameters of the dynamical pictures from* ***Figure*** *5, and the olive dots mark the values of the parameters of the dynamical pictures from* ***Figure*** *6.*

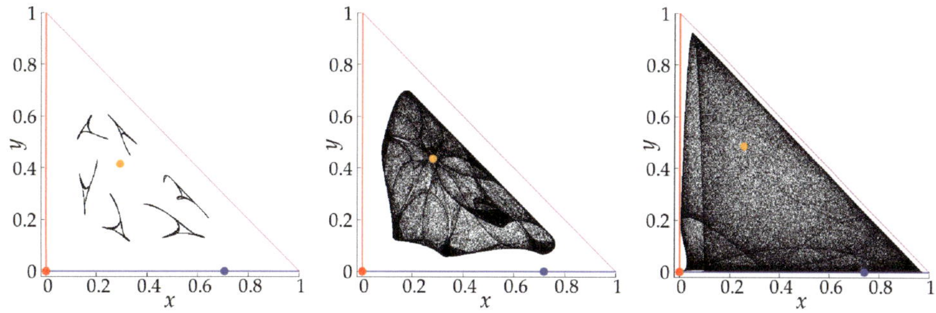

Figure 5.
Plots of the set $\cap_{i=0}^{\infty} T^i(S)$ for $\beta = \mu = 3.412$ (left picture), $\beta = \mu = 3.5485$ (centre picture), and $\beta = \mu = 3.895$ (right picture). In ***Figure 4*** *we can see the location in the parameter space that corresponds to these three dynamical pictures.*

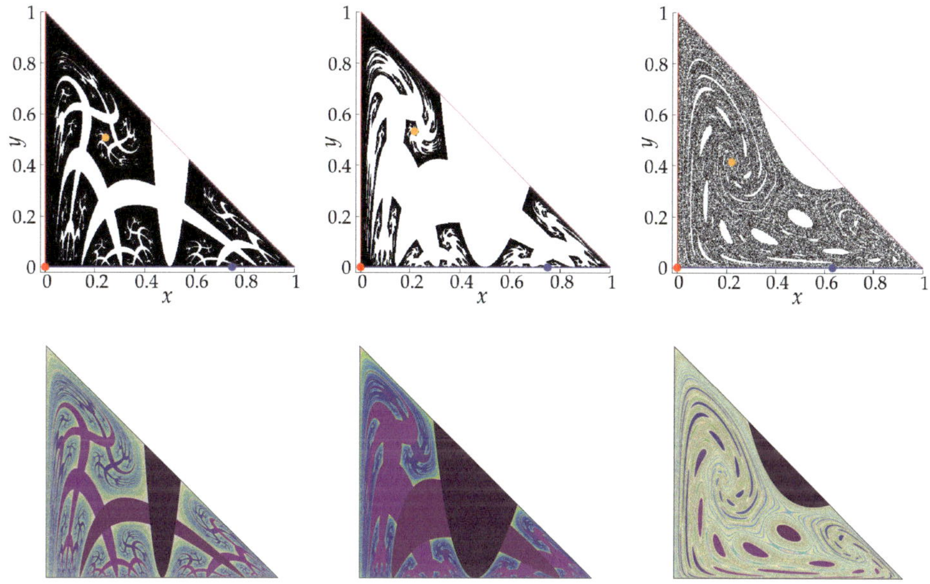

$\beta = 4.1$ and $\mu = 4$ ($\mu^* \approx 3.414286583662$)

$\beta = 4.6$ and $\mu = 4$ ($\mu^* \approx 2.681512445211$)

$\beta = 4.6$ and $\mu = 2.72$ ($\mu^* \approx 2.681512445211$)

Figure 6.
(Top row) The invariant set $S \backslash s\mathcal{R}_{\mu,\beta}$ for several values of $\beta > 4$ and $\mu > \mu^*(\beta)$. In ***Figure 4*** *we can see the location in the parameter space that corresponds to these three dynamical pictures. (Bottom row) Escaping regions with the number of iterates needed to go out of the domain represented in a gradient from 1 (black), 2 (dark violet), 3 (light violet) to 50 (yellow) iterates. Note the fractal nature of the invariant set and of the escaping regions (see movie1.mp4 for an animation of the invariant and escaping sets as a function of model parameters).*

Lemma 4.1.7 from Ref. [13]). Moreover, both sets $\Omega(T)$ and $\cap_{i=0}^{\infty} T^i(S)$ are closed and invariant. Then, in the situation of the above proposition (especially in the light of the above remark), we have $\Omega(T) \subset \cap_{i=0}^{\infty} T^i(S) \not\subseteq S$. To understand the recurrent dynamics of (S, T), it is clearly interesting (and possible) to characterise the set $\cap_{i=0}^{\infty} T^i(S)$ (see **Figure 9** for some examples for different parameter values).

Of course, as we have already implicitly said, in the region at the left and below the magenta curve (see **Figure 4**), one only can expect that $\cap_{i=0}^{\infty} T^i(S)$ will be either $\{P_1^*\}$ or $\{P_1^*, P_2^*\}$, and, hence, it does not draw much attention.

For $\beta > 4$ and $\mu > \mu^*(\beta)$, we also want to characterise the invariant set where the dynamics occur. To this end, we define the *escaping set* $\mathcal{R}_{\mu,\beta}$ as the set of points $z \in S$ such that $T^n_{\mu,\beta}(z) \notin S$ for some $n \geq 1$. Clearly,

$$\mathcal{R}_{\mu,\beta} = \bigcup_{n=0}^{\infty} \left(S \cap T^{-n}_{\mu,\beta}(\varepsilon_{\mu,\beta}) \right) = S \cap \left(\bigcup_{n=0}^{\infty} T^{-n}_{\mu,\beta}(\varepsilon_{\mu,\beta}) \right).$$

As **Figure 6** shows, the set $S \backslash \mathcal{R}_{\mu,\beta}$ is (not surprisingly) much more complicated than the sets S and $S \backslash \varepsilon_{\mu,\beta}$. This prevents obtaining an analytic characterisation of it, as the one given in Proposition 1.11 for the set $S \backslash \varepsilon_{\mu,\beta}$. However, it is always possible (and easy) to obtain numerical approximations to this set for $\beta > 4$ and $\mu > \mu^*(\beta)$ to gain insight about its shape and topology. Observe (see **Figure 6**) that the invariant set $S \backslash \mathcal{R}_{\mu,\beta}$ can be fractal.

Remark 1.10. From the proof of Proposition 1.9, it follows that $\mu^*(\beta)$ is the unique root in the interval $(0, 4)$ of the cubic equation:

$$\mu^3 + \frac{\alpha_2(b)}{\alpha_3(b)}\mu^2 + \frac{\alpha_1(b)}{\alpha_3(b)}\mu + \frac{\alpha_0(b)}{\alpha_3(b)} = 0$$

with $b = \beta - 4$ and

$$\alpha_3(b) = b^2$$

$$\alpha_2(b) = -2(b^3 + 8b^2 + 16b + 32)$$

$$\alpha_1(b) = b^4 + 16b^3 + 96b^2 + 320b + 512, \text{ and}$$

$$\alpha_0(b) = -64(b^2 + 8b + 16).$$

By means of the *Tschirnhaus transformation*

$$\mu = z - \frac{\alpha_2(b)}{3\alpha_3(b)} = z + \frac{2}{3b^2}(b^3 + 8b^2 + 16b + 32),$$

the above equation can be transformed into the following equivalent reduced form:

$$z^3 - \frac{p}{3b^4}z + \frac{2q}{27b^6} = 0 \qquad (9)$$

with

$$\begin{cases} p = -3b^4\left(\dfrac{\alpha_1(b)}{\alpha_3(b)} - \dfrac{\alpha_2(b)^2}{3\alpha_3(b)^2}\right) \\ \quad = b^6 + 16b^5 + 96b^4 + 320b^3 + 1536b^2 + 4096b + 4096, \text{ and} \\ q = \dfrac{27b^6}{2}\left(\dfrac{\alpha_0(b)}{\alpha_3(b)} - \dfrac{\alpha_2(b)\alpha_1(b)}{3\alpha_3(b)^2} + \dfrac{2\alpha_2(b)^3}{27\alpha_3(b)^3}\right) \\ \quad = b^9 + 24b^8 + 240b^7 + 512b^6 - 3840b^5 - 26112b^4 - 88064b^3 - 245760b^2 - 393216b - 262144. \end{cases}$$

Since the linear coefficient of Eq. (9) is negative, it has three real roots, and, by using the trigonometric solution formula for three real root cases, we obtain

$$z^* = 2\sqrt{\frac{1}{3}\frac{p}{3b^4}}\cos\left(\frac{\arccos\left(\left(3\frac{2q}{27b^6}\right)\left(-\frac{1}{2}\frac{3b^4}{p}\right)\sqrt{3\frac{3b^4}{p}}\right)}{3} - \frac{4}{3}\pi\right)$$

$$= \frac{2}{3b^2}\sqrt{p}\cos\left(\frac{\pi - \arccos\left(\frac{q}{p\sqrt{p}}\right)}{3} - \frac{4}{3}\pi\right) = -\frac{2}{3b^2}\sqrt{p}\cos\left(\frac{\arccos\left(\frac{q}{p\sqrt{p}}\right)}{3}\right),$$

and

$$\mu^*(\beta) = z^* - \frac{\alpha_2(b)}{3\alpha_3(b)} = \frac{2}{3b^2}\left(-\sqrt{p}\cos\left(\frac{\arccos\left(\frac{q}{p\sqrt{p}}\right)}{3}\right) + b^3 + 8b^2 + 16b + 32\right).$$

To prove Proposition 1.9, we need a full characterisation of the *one-step escaping set* when $\beta > 4$. This will be obtained in the next proposition.

Proposition 1.11. For every $\beta > 4$,

$$\varepsilon_{\mu,\beta} = \left\{(x,y) : \left|x - \frac{1}{2}\right| < \sqrt{\frac{1}{4} - \frac{1}{\beta}} \text{ and } \frac{1-\mu x(1-x)}{(\beta-\mu)x} < y \leq 1 - x\right\} \neq 0$$

(see **Figure** 7).

Remark 1.12. Observe that $(x,y) \in T^{-1}_{\mu,\beta}(\{(x,y) \in \mathbb{R}^+ \times \mathbb{R}^+ : x + y = 1\})$ if and only if $\mu x(1-x-y) + \beta xy = 1$ which, in turn, is equivalent to

$$y = \frac{1-\mu x(1-x)}{(\beta-\mu)x}.$$

Consequently,

$$\left\{(x,y) \in \mathbb{R}^+ \times \mathbb{R}^+ : y = \frac{1-\mu x(1-x)}{(\beta-\mu)x}\right\} = T^{-1}_{\mu,\beta}(\{(x,y) \in \mathbb{R}^+ \times \mathbb{R}^+ : x + y = 1\})$$

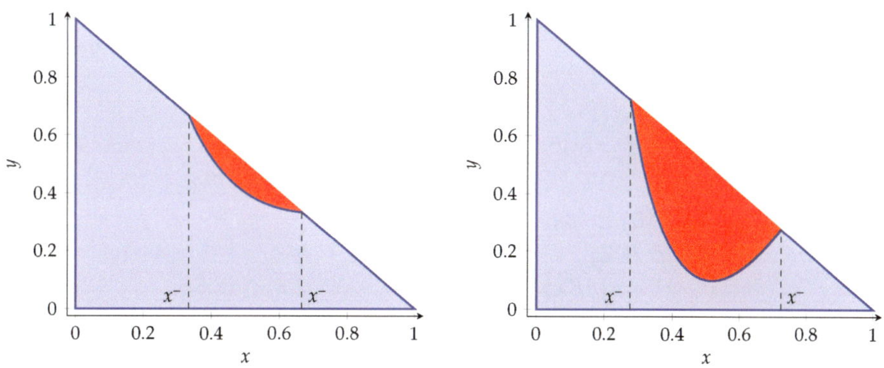

Figure 7.
Two examples of the domain S in blue with the one-step escaping set $\varepsilon_{\mu,\beta}$ plotted in red for $\beta = 4.5$ and $\mu = 2$ (left picture) and $\beta = 5$ and $\mu = 3.75$ (right picture). The one-step escaping set $\varepsilon_{\mu,\beta}$ is vertically delimited by the curves $\frac{1-\mu x(1-x)}{(\beta-\mu)x} < 1 - x$ on the interval with endpoints $x^{\pm} := \frac{1}{2} \pm \sqrt{\frac{1}{4} - \frac{1}{\beta}}$.

and, hence, $\varepsilon_{\mu,\beta}$ is the set of points (x,y) with $\left|x-\frac{1}{2}\right|<\sqrt{\frac{1}{4}-\frac{1}{\beta}}$ which are between the line $u+v=1$ and its $T_{\mu,\beta}$ preimage (in particular they belong to S).

Proof of Proposition 1.11: By assumption we have $\beta>4\geq\mu$. So, additionally, we have $\beta-\mu>0$. We denote

$$x^{-}:=\frac{1}{2}-\sqrt{\frac{1}{4}-\frac{1}{\beta}}\quad\text{and}\quad x^{+}:=\frac{1}{2}+\sqrt{\frac{1}{4}-\frac{1}{\beta}}$$

so that $\left|x-\frac{1}{2}\right|<\sqrt{\frac{1}{4}-\frac{1}{\beta}}$ is equivalent to $x\in(x^{-},x^{+})$. Thus, since

$$0<\frac{\sqrt{5}-1}{2\sqrt{5}}\leq x^{-}<\frac{1}{2}<x^{+}\leq\frac{\sqrt{5}+1}{2\sqrt{5}}<1,$$

$\left|x-\frac{1}{2}\right|<\sqrt{\frac{1}{4}-\frac{1}{\beta}}$ implies $x\in(0,1)$. Hence, $\mu x(1-x)\leq 1$ and $\frac{1-\mu x(1-x)}{(\beta-\mu)x}$ are well defined and non-negative.

To simplify the notation and arguments in the proof, we denote

$$\begin{aligned}\mathrm{E}_{\mu,\beta}&:=\left\{(x,y):\left|x-\frac{1}{2}\right|<\sqrt{\frac{1}{4}-\frac{1}{\beta}}\ \text{and}\ \frac{1-\mu x(1-x)}{(\beta-\mu)x}<y\leq 1-x\right\}\\&=\left\{(x,y):x\in(x^{-},x^{+})\ \text{and}\ \frac{1-\mu x(1-x)}{(\beta-\mu)x}<y\leq 1-x\right\}.\end{aligned}$$

Then, the proposition states that $\mathrm{E}_{\mu,\beta}\neq\emptyset$ and $\varepsilon_{\mu,\beta}=\mathrm{E}_{\mu,\beta}$.

We start by proving that

$$\frac{1-\mu x(1-x)}{(\beta-\mu)x}<1-x\quad\text{if and only if}\quad x\in(x^{-},x^{+}),\tag{10}$$

which implies that the set $\mathrm{E}_{\mu,\beta}$ is a non-empty subset of S, because $(x^{-},x^{+})\subset(0,1)$ and $0\leq\frac{1-\mu x(1-x)}{(\beta-\mu)x}$. To prove (10) observe that

$$\frac{1-\mu x(1-x)}{(\beta-\mu)x}=1-x\Leftrightarrow\frac{1-\beta x(1-x)}{(\beta-\mu)x}=0\Leftrightarrow\beta x(1-x)=1.$$

On the other hand, x^{-} and x^{+} are the two solutions of the equation $\beta x(1-x)=1$. Hence, $\frac{1-\mu x(1-x)}{(\beta-\mu)x}=1-x$ if and only if $x\in\{x^{-},x^{+}\}$. Moreover,

$$\left.\frac{1-\mu x(1-x)}{(\beta-\mu)x}\right|_{x=\frac{1}{2}}=\frac{4-\mu}{2(\beta-\mu)}<\frac{1}{2}=1-\frac{1}{2}$$

because $\beta>4$. So, (10) holds because $\frac{1}{2}\in(x^{-},x^{+})$.

Next we will show that $\mathrm{E}_{\mu,\beta}\subset\varepsilon_{\mu,\beta}$. For every $(x,y)\in\mathrm{E}_{\mu,\beta}\subset\mathrm{S}$, we have $T_{\mu,\beta}(x,y)=(\mu x((1-y)-x),\ \beta xy)$ with $\mu x(1-y-x),\ \beta xy\geq 0$. So,

$$\begin{aligned}\mu x(1-y-x)+\beta xy&=\mu x(1-x)+(\beta-\mu)xy>\mu x(1-x)\\&+(\beta-\mu)x\frac{1-\mu x(1-x)}{(\beta-\mu)x}=1\end{aligned}.$$

Consequently, $T_{\mu,\beta}(x,y)\notin\mathrm{S}$, and hence $(x,y)\in\varepsilon_{\mu,\beta}$.

To end the proof of the lemma, we show the other inclusion: $\varepsilon_{\mu,\beta} \subset E_{\mu,\beta}$, which is equivalent to $S \backslash E_{\mu,\beta} \subset S \backslash \varepsilon_{\mu,\beta}$. From above (see again **Figure** 7) and the fact that for every point $(x,y) \in S$ we have $\mu x(1-y-x)$, $\beta xy \geq 0$, the inclusion $S \backslash E_{\mu,\beta} \subset S \backslash \varepsilon_{\mu,\beta}$ can be written as

$$\begin{aligned}&\{(x,y) : x \in [0,1] \backslash (x^-, x^+) \text{ and } 0 \leq y \leq 1-x\} \bigcup \\ &\left\{(x,y) : x \in (x^-, x^+) \text{ and } 0 \leq y \leq \frac{1-\mu x(1-x)}{(\beta-\mu)x}\right\} = S \backslash E_{\mu,\beta} \subset S \backslash \varepsilon_{\mu,\beta} = \\ &\{z \in S : T_{\mu,\beta}(z) \in S\} = \{(x,y) \in S : \mu x(1-y-x) + \beta xy \leq 1\}.\end{aligned}$$

Let us first consider a point (x,y) such that $x \in [0,1] \backslash (x^-, x^+)$ and $y \in [0, 1-x]$. Since x^- and x^+ are the two solutions of the equation $\beta x(1-x) = 1$, it follows that $x \in [0,1] \backslash (x^-, x^+)$ is equivalent to $\beta x(1-x) \leq 1$. Thus, $\beta > \mu$ gives

$$\mu x(1-y-x) + \beta xy \leq \beta x(1-y-x) + \beta xy = \beta x(1-x) \leq 1.$$

Now we consider a point (x,y) such that $x \in (x^-, x^+)$ and $0 \leq y \leq \frac{1-\mu x(1-x)}{(\beta-\mu)x}$. In this case, in a similar way as before, we have

$$\begin{aligned}\mu x(1-y-x) + \beta xy &= \mu x(1-x) + (\beta-\mu)xy \\ &\leq \mu x(1-x) + (\beta-\mu)x \frac{1-\mu x(1-x)}{(\beta-\mu)x} = 1.\end{aligned}$$

Proof of Proposition 1.9: We will use the characterisation of the set $\varepsilon_{\mu,\beta}$ given by Proposition 1.11. We start by showing the existence of $\mu^* = \mu^*(\beta)$.

Fix $\beta > 4$. Clearly, the parabola $y = \frac{1-\mu x(1-x)}{(\beta-\mu)x}$ and the line $\frac{x}{\mu} + \frac{y}{\beta} = \frac{1}{4}$ intersect if and only if

$$\frac{1-\mu x(1-x)}{(\beta-\mu)x} - \left(\frac{\beta}{4} - x\frac{\beta}{\mu}\right) = 0$$

for some $x \subset \mathbb{R}^+$. This equation is equivalent to

$$\frac{(4\mu^2 + 4\beta(\beta-\mu))x^2 - (4\mu^2 + \beta\mu(\beta-\mu))x + 4\mu}{4\mu(\beta-\mu)x} = 0$$

which, in turn, is equivalent to

$$(4\mu^2 + 4\beta(\beta-\mu))x^2 - (4\mu^2 + \beta\mu(\beta-\mu))x + 4\mu = 0.$$

Thus, the parabola $y = \frac{1-\mu x(1-x)}{(\beta-\mu)x}$ and the line $\frac{x}{\mu} + \frac{y}{\beta} = \frac{1}{4}$ intersect at a unique point if and only if the discriminant of the above quadratic equation is zero:

$$\begin{aligned}0 &= \left(4\mu^2 + \beta\mu(\beta-\mu)\right)^2 - 16\mu\left(4\mu^2 + 4\beta(\beta-\mu)\right) \\ &= \mu\left((\beta(\beta-8)+16)\mu^3 - 2\left(\beta^2(\beta-4)+32\right)\mu^2 + \beta\left(\beta^3+64\right)\mu - 64\beta^2\right).\end{aligned}$$

We need to study the polynomial

$$\begin{aligned}\tilde{P}_0(\mu) &:= (\beta(\beta-8)+16)\mu^3 - 2\left(\beta^2(\beta-4)+32\right)\mu^2 + \beta\left(\beta^3+64\right)\mu - 64\beta^2 \\ &= \alpha_3(b)\mu^3 + \alpha_2(b)\mu^2 + \alpha_1(b)\mu + \alpha_0(b),\end{aligned}$$

where the coefficients $\alpha_i(b)$, with the change of variables $\beta = 4 + b$ with $b \in (0, 1]$, are

$$\alpha_3(b) := \beta(\beta - 8) + 16 = b^2 > 0$$

$$\alpha_2(b) := -2\left(\beta^2(\beta - 4) + 32\right) = -2\left(b^3 + 8b^2 + 16b + 32\right) < 0$$

$$\alpha_1(b) := \beta\left(\beta^3 + 64\right) = b^4 + 16b^3 + 96b^2 + 320b + 512 > 0$$

$$\alpha_0(b) := -64\beta^2 = -64\left(b^2 + 8b + 16\right) < 0.$$

To do it we consider the following sequence of polynomials:

$$\begin{aligned}
P_0(\mu) &:= \frac{\tilde{P}_0(\mu)}{\alpha_3(b)} = \mu^3 + \frac{\alpha_2(b)}{\alpha_3(b)}\mu^2 + \frac{\alpha_1(b)}{\alpha_3(b)}\mu + \frac{\alpha_0(b)}{\alpha_3(b)},\\
P_1(\mu) &:= \frac{1}{3}\frac{\partial P_0(\mu)}{\partial \mu} = \mu^2 + \frac{2\alpha_2(b)}{3\alpha_3(b)}\mu + \frac{\alpha_1(b)}{3\alpha_3(b)},\\
P_2(\mu) &:= -9\alpha_3(b)\mathfrak{R}\mathrm{em}\left(P_0(\mu), P_1(\mu)\right)\\
&= -\left(6\alpha_1(b)\alpha_3(b) - 2\alpha_2(b)^2\right)\mu - 9\alpha_0(b)\alpha_3(b) + \alpha_1(b)\alpha_2(b), \text{ and}\\
P_3(\mu) &:= P_3 := -\mathfrak{R}\mathrm{em}\left(P_1(\mu), P_2(\mu)\right) =\\
&-\frac{81\alpha_0(b)^2\alpha_3(b)^2 + \left(12\alpha_1(b)^3 - 54\alpha_0(b)\alpha_1(b)\alpha_2(b)\right)\alpha_3(b) + 12\alpha_0(b)\alpha_2(b)^3 - 3\alpha_1(b)^2\alpha_2(b)^2}{36\alpha_1(b)^2\alpha_3(b)^2 - 24\alpha_1(b)\alpha_2(b)^2\alpha_3(b) + 4\alpha_2(b)^4} =\\
&b\frac{\begin{array}{l}192b^{10} + 6912b^9 + 113664b^8 + 1069056b^7 + 6438912b^6 +\\ 27131904b^5 + 86507520b^4 + 214695936b^3 + 383778816b^2 + 415236096b + 201326592\end{array}}{\begin{array}{r}b^{12} + 32b^{11} + 448b^{10} + 3712b^9 + 22528b^8 + 118784b^7 +\\ 536576b^6 + 1900544b^5 + 5767168b^4 + 15204352b^3 +\\ 29360128b^2 + 33554432b + 16777216\end{array}} > 0
\end{aligned}$$

where $\mathfrak{R}\mathrm{em}(P, Q)$ denotes the remainder of the division of P by Q (i.e. P modulo Q). Since $P_3(\mu) \neq 0$ for every b, it follows that $\gcd(P_0(\mu), P_1(\mu)) = 1$, and hence $P_0(\mu)$ and $P_1(\mu)$ do not have common roots. In other words, all roots of $P_0(\mu)$ are simple. Consequently, since $\alpha_3(b) > 0$ for every b, the equation $\tilde{P}_0(\mu) = 0$ is equivalent to $P_0(\mu) = 0$, and the above sequence is a Sturm sequence for the polynomial $P_0(\mu)$. The following formulae show this Sturm sequence evaluated at $\mu = 0$ and $\mu = 4$, and the signs of these values:

$$P_0(0) = \frac{\alpha_0(b)}{\alpha_3(b)} < 0,$$

$$P_1(0) = \frac{\alpha_1(b)}{3\alpha_3(b)} > 0,$$

$$P_2(0) = -9\alpha_0(b)\alpha_3(b) + \alpha_1(b)\alpha_2(b) = -2\left(b^7 + 24b^6 + 240b^5 + 1088b^4\right.$$

$$\left. + 2816b^3 + 7680b^2 + 18432b + 16384\right) < 0,$$

$$P_0(4) = 64 + 16\frac{\alpha_2(b)}{\alpha_3(b)} + 4\frac{\alpha_1(b)}{\alpha_3(b)} + \frac{\alpha_0(b)}{\alpha_3(b)} = \frac{64b^2 + 16\alpha_2(b) + 4\alpha_1(b) + \alpha_0(b)}{b^2}$$

$$= \frac{4b^3 + 32b^2 + 128b + 256}{b} > 0,$$

$$P_1(4) = 16 + 4\frac{2\alpha_2(b)}{3\alpha_3(b)} + \frac{\alpha_1(b)}{3\alpha_3(b)} = \frac{48b^2 + 8\alpha_2(b) + \alpha_1(b)}{3b^2} = \frac{b^3 + 16b + 64}{3b} > 0,$$

$$
\begin{aligned}
P_2(4) &= -4\left(6\alpha_1(b)\alpha_3(b) - 2\alpha_2(b)^2\right) - 9\alpha_0(b)\alpha_3(b) + \alpha_1(b)\alpha_2(b) \\
&= -b^2(24\alpha_1(b) + 9\alpha_0(b)) + \alpha_2(b)(\alpha_1(b) + 8\alpha_2(b)) \\
&= -2b\left(b^6 + 20b^5 + 176b^4 + 704b^3 + 1536b(b+1) + 2048\right) < 0, \text{ and} \\
P_3(4) &= P_3 > 0.
\end{aligned}
$$

So, the sign sequences of the Sturm sequence evaluated at $\mu = 0$ and $\mu = 4$ are $\{P_0(0), P_1(0), P_2(0), P_3(0)\} = \{-, +, -, +\}$ which has three changes of sign and $\{P_0(4), P_1(4), P_2(4), P_3(4)\} = \{+, +, -, +\}$ which has two changes of sign. Consequently, the polynomial $P_0(\mu)$ (and hence the polynomial $\tilde{P}_0(\mu)$ and in turn the above discriminant) has a unique root $\mu^* = \mu^*(\beta) \in (0, 4)$. Moreover, since $P_0(0) < 0$ and $P_0(4) > 0$, the discriminant is negative for every $\mu \in (0, \mu^*)$ and positive for every $\mu \in (\mu^*, 4]$. This implies that the parabola $y = \frac{1-\mu x(1-x)}{(\beta-\mu)x}$ and the line $\frac{x}{\mu} + \frac{y}{\beta} = \frac{1}{4}$ do not intersect whenever $\mu < \mu^*$ and intersect at a unique point when $\mu = \mu^*$ (see **Figure 3**). Moreover, for μ small enough and an arbitrary $x \in (0, 1]$, we have

$$
\frac{1-\mu x(1-x)}{(\beta-\mu)x} - \left(\frac{\beta}{4} - x\frac{\beta}{\mu}\right) > 0.
$$

Consequently, since the parabola $y = \frac{1-\mu x(1-x)}{(\beta-\mu)x}$ and the line $\frac{x}{\mu} + \frac{y}{\beta} = \frac{1}{4}$ do not intersect for $\mu < \mu^*$, it follows that

$$
\frac{1-\mu x(1-x)}{(\beta-\mu)x} > \left(\frac{\beta}{4} - x\frac{\beta}{\mu}\right) \quad \text{and} \quad \frac{1-\mu^* x(1-x)}{(\beta-\mu^*)x} \geq \left(\frac{\beta}{4} - x\frac{\beta}{\mu^*}\right)
$$

for every $\beta > 4$, $\mu < \mu^*$, and $x \in (0, 1]$. On the other hand, by Proposition 1.11, the one-step escaping set $\varepsilon_{\mu,\beta}$ is above the parabola $y = \frac{1-\mu x(1-x)}{(\beta-\mu)x}$ and, by definition, it is contained in S. Consequently, for every $\beta > 4$ and $\mu \leq \mu^*$,

$$
\left\{(x,y) \in \mathbb{R}^+ \times \mathbb{R}^+ : \frac{x}{\mu} + \frac{y}{\beta} \leq \frac{1}{4}\right\} \cap \varepsilon_{\mu,\beta} = \emptyset,
$$

which is equivalent to

$$
\mathrm{S} \cap \left\{(x,y) \in \mathbb{R}^+ \times \mathbb{R}^+ : \frac{x}{\mu} + \frac{y}{\beta} \leq \frac{1}{4}\right\} \subset \mathrm{S} \backslash \varepsilon_{\mu,\beta}.
$$

On the other hand, for every $\beta > 4 \geq \mu$, $T_{\mu,\beta}(\mathrm{S}\backslash\varepsilon_{\mu,\beta}) \subset T_{\mu,\beta}(\mathrm{S})$ and, by definition, $T_{\mu,\beta}(\mathrm{S}\backslash\varepsilon_{\mu,\beta}) \subset \mathrm{S}$. Then, by Proposition 1.5, for every $\beta > 4$ and $\mu \leq \mu^*$,

$$
T_{\mu,\beta}(\mathrm{S}\backslash\varepsilon_{\mu,\beta}) \subset \mathrm{S} \cap T_{\mu,\beta}(\mathrm{S}) = \mathrm{S} \cap \left\{(x,y) \in \mathbb{R}^+ \times \mathbb{R}^+ : \frac{x}{\mu} + \frac{y}{\beta} \leq \frac{1}{4}\right\} \subset \mathrm{S} \backslash \varepsilon_{\mu,\beta}.
$$

This proves that the set $\mathrm{S}\backslash\varepsilon_{\mu,\beta}$ is $T_{\mu,\beta}$-invariant for every $\beta > 4$ and $\mu \leq \mu^*(\beta)$.

5. Global dynamics for low values of μ

In this section we investigate global dynamics of the fixed points for low prey's growth rates. This will be done in the next two theorems. In the first one, we show that the fixed point P_1^* is globally asymptotically stable when the intrinsic growth rate of the preys is smaller than 1. See **Figure 4** for a view of these results in the parameter space.

Theorem 1.13 (Global asymptotic stability for $\mu < 1$) We have

$$\lim_{n\to\infty} T^n(x,y) = (0,0) = P_1^*$$

for every $(x,y) \in S\backslash\mathcal{R}_{\mu,\beta}$ (the non-escaping set of T) and $\mu \in (0,1)$.

The proof of this theorem goes "mutatis mutandis" along the same lines as the proof of **Theorem 15** from Ref. [14] by using that, by Lemmas 1.2, 1.3, and 1.4, $P_1^* = (0,0)$ is the unique fixed point of T in S when $\mu \in (0,1)$.

We define

$$\varphi(x) := \begin{cases} 2 & \text{for} x \in [0,2], \\ \dfrac{x}{x-1} & \text{for} x \in [2,5], \end{cases}$$

a continuous non-increasing map from $[0,5]$ to $\left[\frac{5}{4},2\right]$.

Theorem1.14 (Global asymptotic stability for $1 < \mu < \varphi(\beta)$) For every parameter point $(\beta,\mu) \in [0,5] \times (1, \varphi(\beta))$ and $(x,y) \in S\backslash\mathcal{R}_{\mu,\beta}$, we have either

$$T^n(x,y) = (0,0) = P_1^* \quad \text{for some } n \geq 0, \text{or}$$
$$\lim_{n\to\infty} T^n(x,y) = \left(1-\mu^{-1},0\right) = P_2^*.$$

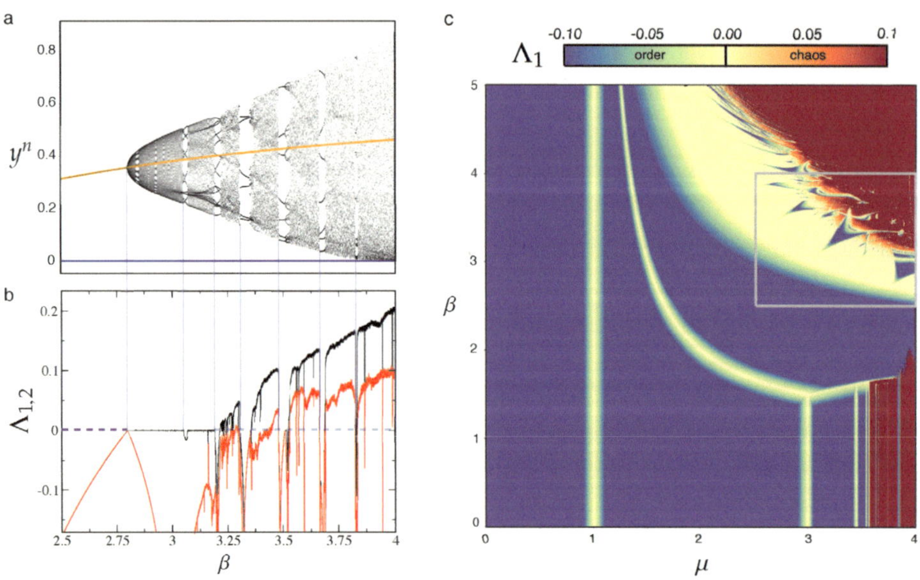

Figure 8.

Bifurcation diagram for Eq. (1) using β as control parameter and $\mu = 2.1$. (a) Dynamics on the attractor for predators at increasing β. The violet and orange lines show the values for fixed points P_2^* and P_3^*, respectively. The vertical blue lines display bifurcations. (b) Spectrum of Lyapunov exponents $\Lambda_{1,2}$, computed for the same range of parameter β. Notice that a Neimark-Sacker bifurcation takes place and the fixed point P_3^* becomes unstable, and after an ordered dynamics with invariant curves and periodic fixed points, the dynamics enters into chaos. The chaotic region displays a wide range of hyperchaos, with two positive Lyapunov exponents. (c) Two-parameter phase diagram displaying the ordered and chaotic dynamics by plotting the first Lyapunov exponent, Λ_1. Note that chaos is found for large values of μ and β (shown in dark yellow-orange-red colours). See movie2.mp4 for an animation of the dynamics of Eq. (1) at increasing both parameters μ and β. The video displays the bifurcation diagram for β and the corresponding attractors obtained numerically.

As before, the proof of this theorem goes "mutatis mutandis" as the proof of Theorem 19 from Ref. [14]) taking into account that, by Lemmas 1.2, 1.3, and 1.4, P_1^* and P_2^* are the unique fixed points of T in S for every $(\beta, \mu) \in [0, 5] \times (1, \varphi(\beta))$. Moreover, P_2^* is the unique locally asymptotically stable fixed point of T in this parameter region. The difference between this theorem and Theorem 19 from Ref. [14] is that, in that paper, β was greater than or equal to 2.5. To recycle the proof of Theorem 19 from Ref. [14] for Theorem 1.14 in the case $\beta \leq 2$, the conditions $1 < \mu < \frac{\beta}{\beta - 1} \leq 2$ and $\alpha_\mu := \frac{\mu - 1}{\mu} < \frac{1}{\beta} \leq \frac{1}{2}$, used in that proof, must be replaced, respectively, by $1 < \mu < \varphi(\beta) = 2$ and $\alpha_\mu < \frac{1}{2} \leq \frac{1}{\beta}$, which play the same role.

6. Chaos

Discrete-time systems can display chaotic behaviour at low dimensions. One example is the well-known logistic model which describes the dynamics of a single species with nonoverlapping generations and intraspecific competition [6]. This system is known to undergo the so-called Feigenbaum (period-doubling) route to chaos [15]. In order to identify the chaotic regions in map (1), we compute the full spectrum of Lyapunov exponents using the algorithm described in Ref. [16], pp. 74–80. **Figure 8(a)** displays a bifurcation diagram obtained by iteration for increasing values of β. Notice that the fixed point P_3^* becomes unstable and a

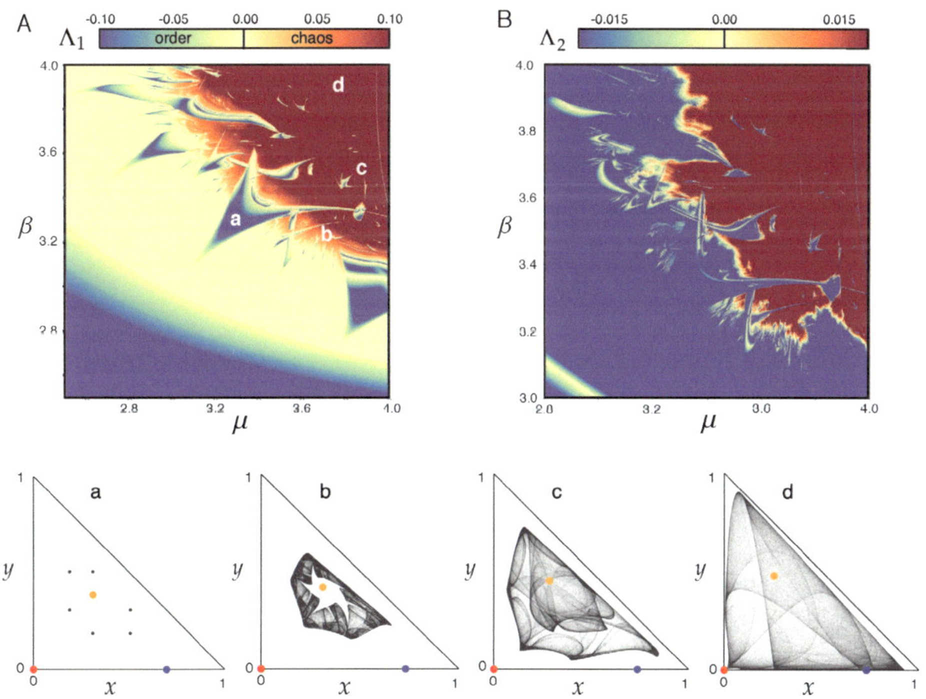

Figure 9.
(A) Enlarged view of the framed region in grey colour in ***Figure 8(c)*** *displaying the first Lyapunov exponent, Λ_1, in the parameter region (μ, $\beta \in [2.5, 4]$). In the orange-red regions, the dynamics are chaotic with $\Lambda_1 > 0$. (B) Second Lyapunov exponent, Λ_2, within the range $\mu \in [2.8, 4]$ and $\beta \in [3, 4]$. The orange-red regions correspond to the hyperchaotic regimes since $\Lambda_{1,2} > 0$. Lower row of pictures: four plots of the set $\cap_{i=0}^{\infty} T^i(S)$ found in the regions labelled with the white numbers in panel (A), period-6 fixed point (a), using $\mu = 3.25$, $\beta = 3.25$, and three examples of strange chaotic attractors, (b) $\mu = 3.7$, $\beta = 3.2$, (c) $\mu = 3.8$, $\beta = 3.5$, and (d) $\mu = 3.7$, $\beta = 3.95$. In all of the phase portraits, we plot the fixed points P_1^* (red), P_2^* (blue), and P_3^* (orange). See movie3.mp4 to visualise the dynamics of Eq. (1) for increasing parameter μ and setting $\beta = 3.9$.*

Neimark-Sacker bifurcation takes place. This bifurcation has been detected with the Lyapunov exponents shown in **Figure 8(b)**, with $\Lambda_{1,2} = 0$ at the bifurcation value. After this bifurcation the first Lyapunov exponent is 0 and the second one is negative. Then the dynamics are governed by attracting invariant curves; further increase of β involves the entry into the chaotic regime, where the first Lyapunov exponent, Λ_1 (in black), becomes positive. Notice the presence of hyperchaotic attractors, with $\Lambda_{1,2} > 0$.

Enlarged views of the Lyapunov exponents in the parameter space (μ, β) are represented in **Figure 9**, as well as four examples of the sets $\cap_{i=0}^{\infty} T^i(S)$ found in the regions labelled with letters in **Figure 9(A)**.

7. Conclusions

In this chapter we have analysed the dynamics of a predator-prey dynamical system in discrete time (see also [12]). We have provided conditions for the global stability of the fixed points corresponding the co-extinctions of the predator-prey as well as for the extinction of predators and survival of preys. For some parameter regions, we have identified hyperchaos (i.e. more than one positive Lyapunov exponent; see [17]). A deep analysis of the existence and properties of the invariant set has been provided for a wide region of the parameter space containing the most biologically relevant dynamics. We have identified the presence of escaping zones in the phase space at which species populations go out of the domain (e.g. they overcome the carrying capacity) and then the iterates become negative, meaning that populations go to extinction. By means of iteration, we have characterised a very complicated shape of the escaping regions, presumably with a highly entangled, fractal topology. These escaping regions could be responsible for species extinctions evolving in discrete time. Although early experimental research allowed to identify deterministic chaos in insect populations [3], as far as we know, no empirical proofs about this phenomenon have been described.

Acknowledgements

We want to thank Ricard Solé, Sergi Valverde, and Tomás Alarcón for useful comments. LlA has been supported by the by Spain's "Agencia Estatal de Investigación" (AEI) grant MTM2017-86795-C3-1-P. JTL has been partially supported by the Catalan grant 2017SGR1049 and the MINECO grant MTM2015-65715-P and PGC2018-098676-B-100 (AEI/FEDER/UE). BV was funded by the PR01018-EC-H2020-FET-Open MADONNA project. This work has been also partially funded by the CERCA Program of the Generalitat de Catalunya and by the MINECO grant MDM-2014-0445 within the "María de Maeztu" Program. JS has been funded by a "Ramón y Cajal" contract RYC-2017-22243 and by the MINECO grant MTM2015-71509-C2-1-R and AEI grant RTI-2018-098322-B100.

Author details

Blai Vidiella[1,2†], J. Tomás Lázaro[3,4†], Lluís Alsedà[5,6,4*] and Josep Sardanyés[6,4*]

1 ICREA-Complex Systems Lab, Universitat Pompeu Fabra, Barcelona, Spain

2 Institut de Biologia Evolutiva, CSIC-Universitat Pompeu Fabra, Barcelona, Spain

3 Departament de Matemàtiques, Universitat Politècnica de Catalunya, Barcelona, Spain

4 Barcelona Graduate School of Mathematics (BGSMath), Bellaterra, Spain

5 Departament de Matemàtiques, Universitat Autònoma de Barcelona, Bellaterra, Spain

6 Centre de Recerca Matemàtica, Bellaterra, Spain

*Address all correspondence to: llalseda@crm.cat and jsardanyes@crm.cat

† Equal contribution.

References

[1] Dennis B, Desharnais RA, Cushing JM, Henson SM, Constantino RF. Estimating chaos and complex dynamics in an insect population. Ecological Monographs. 2001;7(12):277-303

[2] Elton CS. Fluctuations in the numbers of animals: their causes and effects. British Journal of Experimental Biology. 1924;**2**:119-163

[3] Constantino RF, Desharnais RA, Cushing JM, Dennis B. Chaotic dynamics in an insect population. Science. 1997;**275**:389-339

[4] Dennis B, Desharnais RA, Cushings JM, Constantino RF. Estimating chaos and complex dynamics in an insect population. The Journal of Animal Ecology. 1997;**66**:704-729

[5] May RM. Biological populations with nonoverlapping generations: Stable points, stable cycles and chaos. Science. 1974;**186**:645-647

[6] May RM. Simple mathematical models with very complicated dynamics. Nature. 1976;**261**:459-467

[7] Allen JC, Schaffer WM, Rosko D. Chaos reduces species extinction by amplifying local population noise. Nature. 1993;**364**:229-232

[8] Davies ZG, Wilson RJ, Brereton TM, Thomas CD. The re-expansion and improving status of the silver-spotted skipper butterfly (*Hesperia comma*) in Britain: a metapopulation success story. Biological Conservation. 2005;**124**: 189-198

[9] Krafsur ES. Gene flow between univoltine and semivoltine northern corn rootworm (Coleoptera: Chrysomelidae) populations. Annals of Enthomological Society of America. 1995;**88**:699-704

[10] Saulich AK, Musolin DL. Seasonal cycles in stink bugs (*Heteroptera, Pentatomidae*) from the temperate zone: Diversity and control. Entomological Review. 2014;**94**:785-814

[11] Aalberg Haugen IM, Berger D, Gotthard K. The evolution of alternative developmental pathways: Footprints of selection on life-history traits in a butterfly. Journal of Evolutionary Biology. 2012;**25**:1388-1388

[12] Lauwerier HA. Two-dimensional iterative maps. In: Arun V, editor. Chaos. Holden: Princeton University Press; 1986. pp. 58-95

[13] Alsedà L, Llibre J, Misiurewicz M. Combinatorial dynamics and entropy in dimension one. In: Volume 5 of Advanced Series in Nonlinear Dynamics. 2nd ed. River Edge, NJ: World Scientific Publishing Co., Inc.; 1989

[14] Alsedà Ll, Vidiella B, Solé R, Lázaro JT, Sardanyés J. Dynamics in a time-discrete food-chain model with strong pressure on preys. Communications in Nonlinear Science and Numerical Simulation. 2020; **84**:105187

[15] Feigenbaum MJ. Universality in complex discrete dynamics. Los Alamos Theoretical Division Annual Report 1975–1976

[16] Parker T, Chua LO. Practical Numerical Algorithms for Chaotic Systems. Berlin: Springer-Verlag; 1989

[17] Li P, Min L, Yu H, Zhao G, Li X. Novel two dimensional discrete chaotic maps and simulations. In: IEEE 6th International Conference on Information and Automation for Sustainability (ICIAFS). 2012

Chapter 8

Using Dynamic Analysis to Adjust the Rheological Model of Three Parameters to the Eurocode Creep Criteria

Alexandre de Macêdo Wahrhaftig

Abstract

A dynamic analysis of vibration for considering a three-parameter rheological model to fit the same results as predicted for creep by the Eurocode (EN 1992) criteria is performed based on the adjustment of its parameters. The use of a rheological model of three parameters as a valid alternative for real problems brings a huge facility for mathematical implementation and manipulation due the simplicity of the solution. For adjustment of the elastics and the viscous parameters, a numerical simulation to calculate the fundamental frequency of an actual reinforcement concrete pole is carried out in comparison with the standard Eurocode criteria. In this determination, the geometry variation, a concentrated force present at the free end of the structural element, and the self-weight of the structure are considered. The physical nonlinearity of the concrete due to the cracks is also considered by reducing of the flexural stiffness, and its viscoelastic behavior is included in the calculation through a temporal modulus of elasticity. In the analysis, the ground was modeled as a set of distributed springs along the foundation length. The frequency over time is then analytically calculated as the critical buckling load for different instants after the structure to be loaded.

Keywords: vibration, analytical solution, Eurocode, viscoelasticity, creep, geometric nonlinearity, material nonlinearity

1. Introduction

A column represents a continuous structural member whose vibrations are governed by nonlinear partial differential equations for which exact analytical solutions cannot be found, as pointed out by [1]. Columns constitute continuous systems, and their analysis can be reduced to an analogous system containing a single degree of freedom. The vibration mode is restricted to a configuration previously established by a mathematical function that describes the vibratory movement, and the properties of the system can be expressed as generalized coordinate functions [2]. In his study on the vibration of elastic systems applied this technique considering the function valid throughout the problem domain. However, for real cases, where the properties of the structural elements vary along their length, the formulation developed for calculating the stiffness and mass must be solved by observing the intervals defined in the geometry. In these cases, the integrals obtained can be solved within the limits established for each interval, i.e., the generalized properties can be calculated for each discrete segment of the structure, as defined by its geometry within that segment. A variety of vibration problems using that mathematical concept were solved by [3] who mentioned a previous one [4] where the buckling load is calculated for stepped and tapered columns and where how laborious or even impossible it is to apply it for problems with variable geometry is registered. With the advent of digital computers, these problems passed to be solved by modeling that use discretization technics of the continuum [5].

To analytically define the fundamental frequency for the case modeled in this study, all the elastic stiffness components are considered in the calculation, including the conventional stiffness, which depends on the material behavior; the geometric stiffness, which depends on the normal force acting on the structure; and the soil parcel, which accounts for the soil-structure interaction. It is important to note that the soil-structure interaction cannot be ignored, particularly in the case of a monopile foundation, because it may significantly influence the dynamic behavior of the structure [6].

The structure selected for this study is a slender reinforced concrete (RC) having both full and hollow circular section with variable geometry, for which the natural frequency and the critical buckling load were calculated considering all nonlinearities present in the system. It is important to highlight that nonlinearities play an important role when calculating dynamic proprieties of a system, as well pointed by [7]. In this work, the geometric nonlinearity was taken in consideration by using the geometric stiffness parcel into the total stiffness of the system. The nonlinearity of the material was computed by reducing its flexural stiffness, as similarly done by [8], reflecting the development of cracking in the concrete when bended, which is dependent on the magnitude of the stress. Another kind of material nonlinearity is creep, which occurs due the viscoelastic behavior of the concrete, it being considered in two ways. The first one is the mathematical model for creep predicted by Eurocode 2 (European Standard EN 1992-1-1) [9]. The second one is a three-parameter viscoelastic model whose parameters are adjusted in order to meet the results obtained when using the Eurocode. In this sense, the use of the three-parameter viscoelastic model to represent the creep of concrete brings an enormous facility of employment for actual cases due the reduced number of variables which are manipulated. Indeed, just one of them is necessary because two of the three parameters can be expressed in terms of the modulus of elasticity of the concrete, a data easily calculated for any standard procedure or obtained in laboratory.

2. Analytical solution of the structural frequency

Figure 1 presents the bar model of a structure in free vibration. Consider the following trigonometric function, taken as valid throughout its domain:

$$\phi(x) = 1 - \cos\left(\frac{\pi x}{2L}\right), \tag{1}$$

where x is the location of the calculation, originating at the base of the cantilever, and L is the length of the column.

That model represents a column under an axial compressive load, $N_s(x)$, with either constant or variable properties along its length. These properties include the geometry, elasticity/viscoelasticity, and density. Applied springs of variable stiffness $k_{so}(x)$ act as the lateral soil resistance until the foundation elevation Gr.

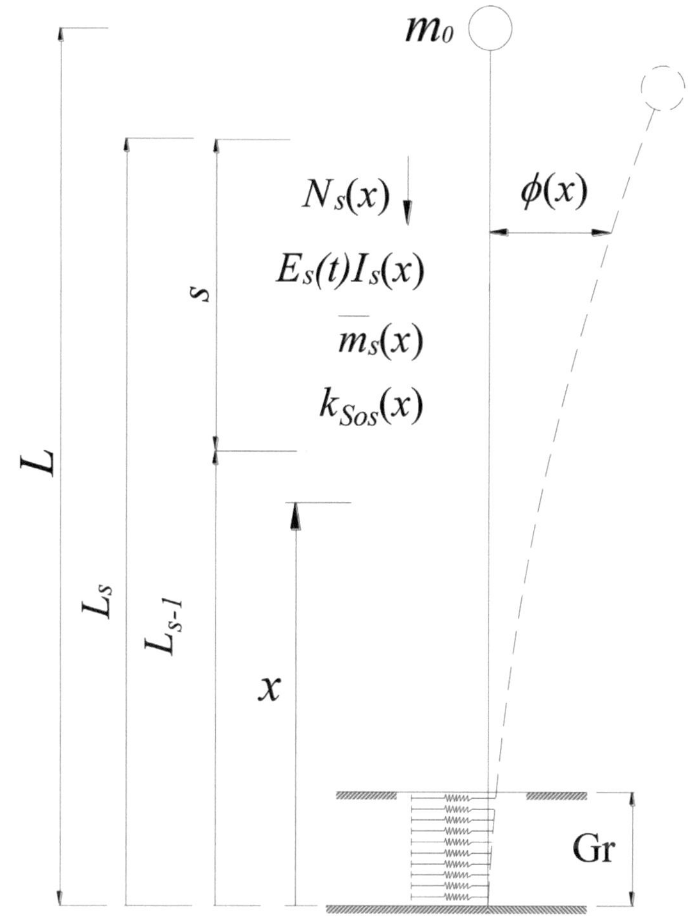

Figure 1.
Frame element model in free vibration.

The system is under the action of gravitational normal forces, originating from the distributed mass along the length of the column and of a lumped mass at the tip m_0.

In the case of vibration of a cantilevered column that is clamped at its base and free at its tip, the shape function given in Eq. (1) satisfies the boundary conditions of the problem. The use of Eq. (1) as a shape function for an actual structure with varying geometry has been validated by [10]. This validation involved a comparison with a computational solution derived using computational modeling by finite element method (FEM) and other mathematical expressions.

By applying the principle of virtual work and its derivations, the dynamic properties of the subject system are obtained. The elastic/viscoelastic conventional stiffness is given by

$$k_{0s}(t) = \int_{L_{s-1}}^{L_s} E_s(t) I_s(x) \left(\frac{d^2\phi(x)}{dx^2} \right)^2 dx, \text{ with } K_0(t) = \sum_{s=1}^{n} k_{0s}(t), \tag{2}$$

where for a segment s of the structure, $E_s(t)$ is the viscoelastic modulus of the material with respect to time; $I_s(x)$ is the variable moment of inertia of the section along the segment in relation to the considered movement, obtained by interpolation of the previous and following sections and if it is constant, it is simply I_s; $k_{0s}(t)$ is the temporal term for the stiffness; $K_0(t)$ is the final conventional stiffness varying over time; and n is the total number of segment intervals given by the structural geometry. In Eq. (2), obviously, t vanishes when the analysis considers a material with purely elastic, time-independent behavior. The geometric stiffness appears as a function of the axial load, including the self-weight contribution and is expressed as

$$k_{gs}(m_o) = \int_{L_{s-1}}^{L_s} \left[N_0(m_o) + \sum_{j=s+1}^{n} N_j + \overline{m}_s(x)(L_s - x)g \right] \left(\frac{d\phi(x)}{dx} \right)^2 dx \text{ and} \tag{3}$$

$$K_g(m_0) = \sum_{s=1}^{n} k_{gs}(m_0), \tag{4}$$

where $k_{gs}(m_0)$ is the geometric stiffness in segment s, $K_g(m_0)$ is the total geometric stiffness of the structure with n as defined previously, and $N_0(m_0)$ is the concentrated force at the top, all of which are dependent on the mass m_0 at the tip, given by

$$N_0(m_0) = m_0 g. \tag{5}$$

Further, N_j is the normal force from the upper segments, obtained by

$$N_j = \int_{L_{s-1}}^{L_s} \overline{m}_s(x) g dx. \tag{6}$$

Then, the total generalized mass is given by

$$M(m_0) = m_0 + m, \tag{7}$$

considering that

$$m = \sum_{s=1}^{n} m_s, \text{ with } m_s = \int_{L_{s-1}}^{L_s} \overline{m}_s(x)(\phi(x))^2 dx, \text{ and } \overline{m}_s(x) = A_s(x)\rho_s, \tag{8}$$

where $\overline{m}_s(x)$ is the mass distributed to each segment s, which is obtained by multiplying the cross-sectional area, $A_s(x)$, by the density, ρ_s, of the material in the respective interval. Therefore, $\overline{m}_s(x)$ is the mass per unit length, and m is the generalized mass of the system owing to the density of the material, with n as previously defined. If the cross section has a constant area over the interval, $A_s(x)$ will be just A_s; consequently, the distributed mass will also be constant. Similarly, if the mass m_0 does not vary, all the other parameters that depend on it will also be constant.

One approach for considering the participation of the soil in the vibration of the system is to consider it as a series of vertically distributed springs that act as a restorative force on the system. With $k_{Sos}(x)$ denoting the spring parameter, the effective soil stiffness (as a function of the location x along the length) is generally defined as

$$K_{So} = \sum_{s=1}^{n} k_s, \text{ with } k_s = \int_{L_{s-1}}^{L_s} k_{Sos}(x)\phi(x)^2 dx, \text{ where } k_{Sos}(x) = S_{os} D_s(x), \tag{9}$$

where the parameter K_{So} is an elastic characteristic consisting of the sum of $k_{Sos}(x)$ along the foundation depth, which depends on the geometry of the foundation $D_s(x)$ and the soil parameter S_{os}. Considering the normal force as positive, the total structural stiffness is obtained as

$$K(t, m_0) = K_0(t) - K_g(m_0) + K_{So}. \tag{10}$$

Finally, the natural frequency (in Hertz), as a function of the time and the mass at the tip, is calculated according to Eq. (11). The great advantage of using that equation in terms of two independent variables is that it can be employed to calculate the critical load of buckling as well, because all the generalized parameters are expressed as a function of the mass at the top. Details of this analytical procedure can be seen in [11]:

$$f(t, m_0) = \frac{1}{2\pi}\sqrt{\frac{K(t, m_0)}{M(m_0)}}. \tag{11}$$

3. Creep consideration

The creep represents the increase of deformation under constant stress, which occurs in some materials due to its viscoelastic nature. It is essential to consider it in the analysis of slender structural elements, because the stiffness of these members is modified as a function of the rheology of the material. Usually, viscoelasticity is associated with creep of structural elements and can be characterized by models where the immediate elastic deformation is increased by viscous deformation, resulting in a temporal function for deformation. Consequently, the modulus of elasticity must also be provided as a temporal function that provides accurate results under normal levels of stress. Due to the viscous nature of the concrete, even at a constant stress level, the deformation of a structural element tends to increase over time. An increase in strain over time under constant stress is a viscoelastic phenomenon.

3.1 Solution of the three-parameter rheological model

Mathematically, viscoelasticity can be represented by a time-dependent function associated with rheological models capable of describing the phenomenon. It is conceptually convenient to consider classic viscoelastic models in which there are only two types of parameters, relating to elasticity and viscosity. Classic viscoelastic models are obtained by arranging springs and dampers, or dashpots, in different configurations. Springs are characterized by elastic moduli and dashpots by viscosity coefficients. The best known of these mechanical models are the Maxwell model, containing a spring in series with a dashpot, and the Kelvin-Voigt model, containing a spring and dashpot in parallel [12]. One model used to represent the viscoelasticity of solids is the three-parameter model, in which the elastic parameter E_e is connected to the viscoelastic Kelvin-Voigt model with parameters E_v and η, which is a simplification of Group I of the Burgers model, as shown in **Figure 2**.

The three-parameter model is an appropriate model for describing the viscoelastic nature of many solids [13] and is often used to study the phenomenon in various scientific fields. Adaptation of the Burgers model in different fields of structural analysis can be found in [14–19]. The total deformations of the Kelvin-Voigt model are given by $\varepsilon = \varepsilon^e + \varepsilon^v$, where ε^e is the deformation of the elastic model

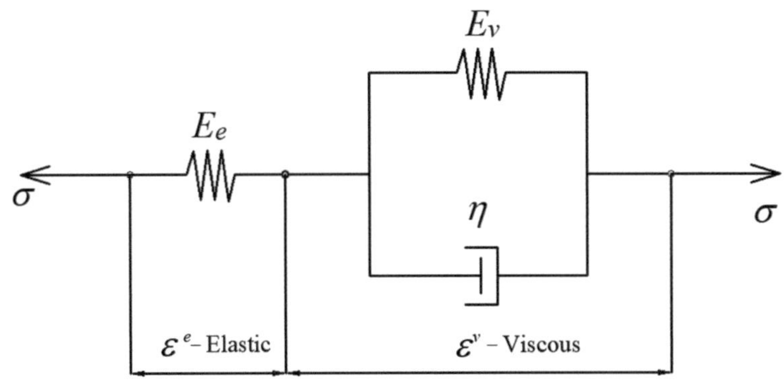

Figure 2.
Viscoelastic model of three parameters.

and ε^v is the deformation of the Kelvin-Voigt model. When differentiated with respect to time, the total deformation is obtained as

$$\dot{\varepsilon} = \dot{\varepsilon^e} + \dot{\varepsilon^v}, \tag{12}$$

which includes the constitutive equations of the elastic and Kelvin-Voigt models, respectively. Considering the modulus of elasticity for both parts, elastic and viscous, the stress becomes

$$\sigma = E_e\varepsilon^e \text{and } \sigma = E_v\varepsilon^v + \eta\, \dot{\varepsilon^v}. \tag{13}$$

From the previous equations, one derives the following differential equation:

$$\dot{\sigma} + \frac{E_e + E_v}{\eta}\sigma = E_e\, \dot{\varepsilon} + \frac{E_eE_v}{\eta}\varepsilon, \tag{14}$$

where $\sigma = 0$ for $t < 0$ and $\sigma = \sigma_0$ for $t > 0$, with t representing the time and $t = 0$ the instant of loading application. As the stress remains constant, the derivative of the stress with respect to time is zero. Applying the previous stress condition, the following ordinary differential equation is found:

$$E_e\, \dot{\varepsilon} + \frac{E_eE_v}{\eta}\varepsilon = \sigma_0, \tag{15}$$

for which the general solution for $t > 0$, taking the initial condition $\varepsilon(0) = \sigma_0/E_e$, is

$$\varepsilon(t) = \sigma_0\left[\frac{1}{E_e} + \frac{1}{E_v}\left(1 - e^{-\frac{E_v}{\eta}t}\right)\right]. \tag{16}$$

From Eq. (16), it is possible to extract the temporal function for the modulus of elasticity of the three-parameter model:

$$E(t) = \frac{1}{\left[\frac{1}{E_e} + \frac{1}{E_v}\left(1 - e^{-\frac{E_v}{\eta}t}\right)\right]}. \tag{17}$$

Therefore, it is easily seen that for

$$t = 0 \Rightarrow e^{-\frac{E_v}{\eta}} = 1 \therefore \varepsilon(0) = \frac{\sigma_0}{E_e}; \tag{18}$$

$$t \to \infty \Rightarrow e^{-\frac{E_v}{\eta}} = 0 \therefore \varepsilon(\infty) = \frac{\sigma_0 (E_e + E_v)}{E_e E_v}, \tag{19}$$

$$E(\infty) = \frac{E_e E_v}{E_e + E_v} \Rightarrow \varepsilon(\infty) = \frac{\sigma_0}{E(\infty)}. \tag{20}$$

It is important to note that the viscoelastic behavior of the considered material is completely represented by the temporal modulus of elasticity and it can be used for static or dynamic applications. For instance, the previous solution was used in numerical simulations as can be seen in [20, 21].

It is possible to transform the parameters of the viscous part to being just a function of the modulus of elasticity of the elastic part, which can easily be calculated by any standard procedure or obtained in the simplest laboratory. Therefore, these parameters can be written as

$$E_v = \alpha E_e; \eta = \gamma E_e, \tag{21}$$

where α is a real positive number and γ brings together a temporal unit.

3.2 Model predicted by Eurocode

The method specified in European Standard EN 1992-1-1 for incorporating creep into structural analysis considers the effects of the creep behavior and its variation with time. Eurocode 2 provides hypothetical and model limitations for creep calculation, wherein the creep coefficient φ is predicted as a function of the tangent modulus of elasticity E_c. The creep deformation of concrete is computed by multiplying the immediate deformation by the creep coefficient. The total concrete deformation at time t, under constant temperature, can be obtained as the sum of the terms that represent the immediate deformation and creep. All the factors related to the phenomenon, such as loading and environment humidity, are calculated under the assumption that they remain constant over the considered time interval, affording a specific result for the creep coefficient φ. This coefficient is then directly introduced into the slow deformation equation and used as input data for various procedures. The basic equations for determining the creep coefficient of concrete over time are based on the average compressive strength f_{cm} ($f_{cm} = f_{ck} + 8$, f_{ck} in MPa). The creep coefficient $\varphi(t, t_0)$, as defined in Eq. (22), is the product of two factors, namely, φ_0 and $\beta_c(t, t_0)$, which, respectively, characterize the effects of the rheological properties of the concrete under environmental conditions and the evolution of creep with time after loading of the structure:

$$\varphi(t, t_0) = \varphi_0 \beta_c(t, t_0). \tag{22}$$

The first factor φ_0 defined in Eq. (23) consists of three other factors. The first of them, φ_{RH} (given by Eq. (24)), considers concrete compressive strengths >35 MPa (as in the case that will be seen) and accounts for the effects of the environmental relative humidity RH, the equivalent thickness h_0 of the member which is a function of the cross-sectional area A_c, and the external perimeter u_e of the member in contact with the environment. The second one, $\beta(f_{cm})$ (Eq. (26)), represents the direct effect of the resistance on φ_0. The third, $\beta(t_0)$ (Eq. (27)), takes into account the age of the concrete at the beginning of loading, i.e., at t_0.

$$\varphi_0 = \varphi_{RH}\beta(f_{cm})\beta(t_0), \tag{23}$$

$$\varphi_{RH} = \left[1 + \frac{1 - RH/100}{0.1\sqrt[3]{h_0}\alpha_1}\right]\alpha_2, \tag{24}$$

$$h_0 = \frac{2A_c}{u_e}, \tag{25}$$

$$\beta(f_{cm}) = \frac{16.8}{\sqrt{f_{cm}}}, \tag{26}$$

$$\beta(t_0) = \frac{1}{(0.1 + t_0{}^{0.20})}. \tag{27}$$

The second factor, $\beta_c(t, t_0)$ (Eq. (28)), is a function of the coefficient β_H (given by Eq. (29) for average concrete compressive strengths upper than 35 MPa), and it is used to regulate the combined effects of the relative humidity and the equivalent member thickness. The percolation path of the adsorbed water in a robust section of concrete is so large that the effects of creep due to differential moisture are less important for slimmer sections.

$$\beta_c(t, t_0) = \left[\frac{(t - t_0)}{\beta_H + (t - t_0)}\right]^{0.3}, \tag{28}$$

$$\beta_H = 1.5\left[1 + (0.012RH)^{18}\right]h_0 + 250\alpha_3 \leq 1500, \tag{29}$$

$$\alpha_1 = \left(\frac{35}{f_{cm}}\right)^{0.7}, \alpha_2 = \left(\frac{35}{f_{cm}}\right)^{0.2}, \alpha_3 = \left(\frac{35}{f_{cm}}\right)^{0.5}. \tag{30}$$

Thus, the creep coefficient can be obtained using Eq. (22), and the temporal function that describes the deformation in accordance with EN 1992-1-1 can be expressed as

$$\varepsilon(t, t_0) = \sigma_c(t_0)\left[\frac{1}{E_c(t_0)} + \frac{\varphi(t, t_0)}{E_c(t_{28})}\right]. \tag{31}$$

Based on the above equations, the modulus of elasticity with respect to time can be expressed as

$$E(t, t_0) = \frac{1}{\frac{1}{E_c(t_0)} + \frac{\varphi(t, t_0)}{E_c(t_{28})}}, \tag{32}$$

where $E_c(t_0)$ is the modulus of elasticity at the beginning of loading and $E_c(t_{28})$ is the modulus of elasticity 28 days after the commencement of loading.

4. An application

The case selected for the present study involves calculating the fundamental frequency and the critical buckling load of an actual slender reinforced concrete pole with variable geometry that presents both geometrical and material nonlinearities as shown in **Figure 3**.

The structure is 46 m high, which includes a 40 m superstructure with a hollow circular section and a 6-m-deep, full circular-type foundation. The moduli of elasticity adopted for the superstructure and foundation are 30.24 and 24.97 GPa, calculated by Eq. (33) considering characteristic resistances (f_{ck} at 28 days after production) of 45 and 20 MPa, respectively:

$$E_e = \frac{22}{1.2}\left(\frac{f_{ck}+8}{10}\right)^{0.3} (\text{GPa}) \ (f_{ck} \text{ in MPa}). \tag{33}$$

A set of antennas and a platform are installed at the tip of the structure, constituting a concentrated mass of 1098 kg. Cables and a ladder are installed along the entire length, adding a distributed mass to the system of 40 kg/m. The densities of the reinforced concrete were defined as 2600 and 2500 kg/m^3 for the super- and infrastructure, respectively. The physical nonlinearity of the material was

Figure 3.
Photographic images of the actual slender reinforced concrete.

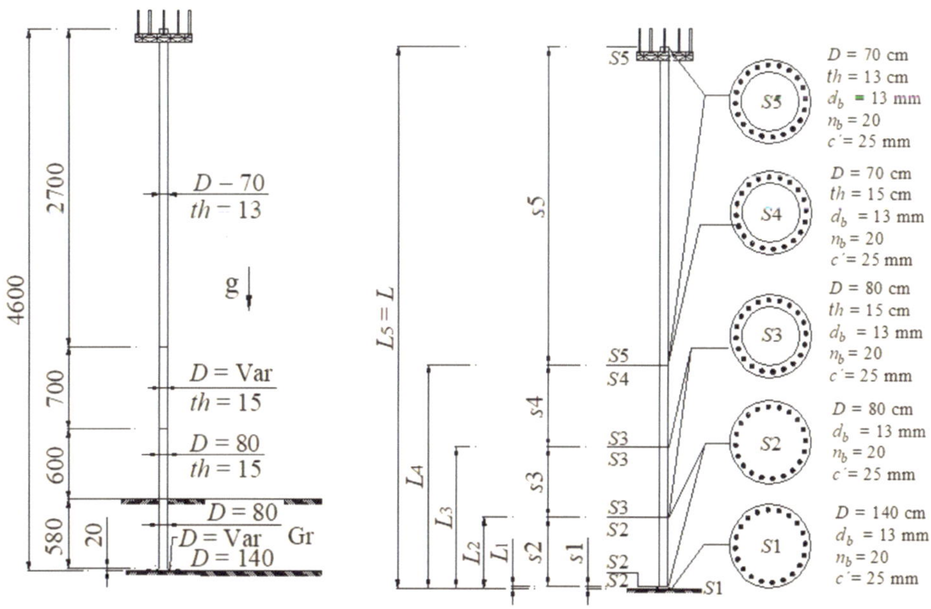

Figure 4.
Subject reinforced concrete pole: (a) geometry (cm); (b) sections.

computed for the superstructure and the foundation reducing the gross moment of inertia by a multiplier factor equal to 0.3, allowing the performing of a simplified nonlinear analysis according to Eurocode 2, as presented in [22], but being possible the use of other coefficients as explained by [23].

The foundation is a relatively deep shaft having a bell diameter of 140 cm, bell length of 20 cm, shaft diameter of 80 cm, and shaft length of 580 cm. The lateral soil resistance is represented by an elastic parameter, S_{os}, equal to 2667 kN/m^3.

The geometric details of the evaluated pole are shown in **Figure 4**, where g denotes gravitational acceleration; Gr means ground; s represents each structural segment; S, D, and th are the type, the external diameter, and the wall thickness of the section; d_b represents the reinforcing bar diameter; n_b is the number of reinforcing bars; and c' is the reinforcing cover. The slenderness ratio of the tower structure is approximately 400.

Because this is an RC structure, it is necessary to account for the presence of the reinforcing bars when calculating the moment of inertia, which is accomplished by homogenizing the cross section. Therefore, according to the theorem of parallel axis, the factors, which multiply the nominal moment of inertia of the section in terms of the total moment of inertia of the reinforcing steel, in the homogenized section are appropriately calculated. Studies that assure the occurrence of the transfer of creep to the reinforcement of columns were development by [24, 25].

5. Simulation results

Considering E_e is equal to $E(t_0) = E(t_{28})$ and setting α and γ as 3.913 and $10^{8.3066}$ seconds, respectively, and adopting an environmental humidity of 70%, the variation of the fundamental frequency for different instants in the lifetime of the structure can be obtained. The produced results by using the Eurocode model can be observed in **Figure 5a**, and the result by using the adjusted three-parameter viscoelastic model can be seen in **Figure 5b**. It is important to mention that these chosen values for α and γ were defined so that simulation leads a good agreement for instants approaching and after 2000 days. Therefore, they were intentionally defined so that the frequency met the same values as given by Eurocode. The choice of these coefficients has been done because the convergence of the deformations occurs at 4000 days, at which time the interest of the structural engineering normally lies, being, however, possible to define other pairs of values for α and γ in the case of a particular objective or even to choose which can match both curves in the whole time interval. Therefore, the mentioned coefficients have been adjusted so that the frequency is equalized by both models considering a precision of six significant digits, as can be highlighted in **Table 1**. When the modulus of elasticity is calculated by both models, that precision is not reached.

Figure 6 shows a comparison between results produced through both models, considering each selected instants of time.

By using the presented dynamic procedure, the critical buckling load is determined when the frequency is zero at any arbitrary time after the structure gets into service. Taking all the previous explanation into consideration and varying the mass at the tip, the force acting at the top also varies according to Eq. (5), as does the frequency of the structure that varies according to Eq. (11). The results obtained for the buckling load for both models can be seen in **Figure** 7. To obtain it, a short routine of programming has been elaborated considering increments of 0.1 kg to the lumped mass.

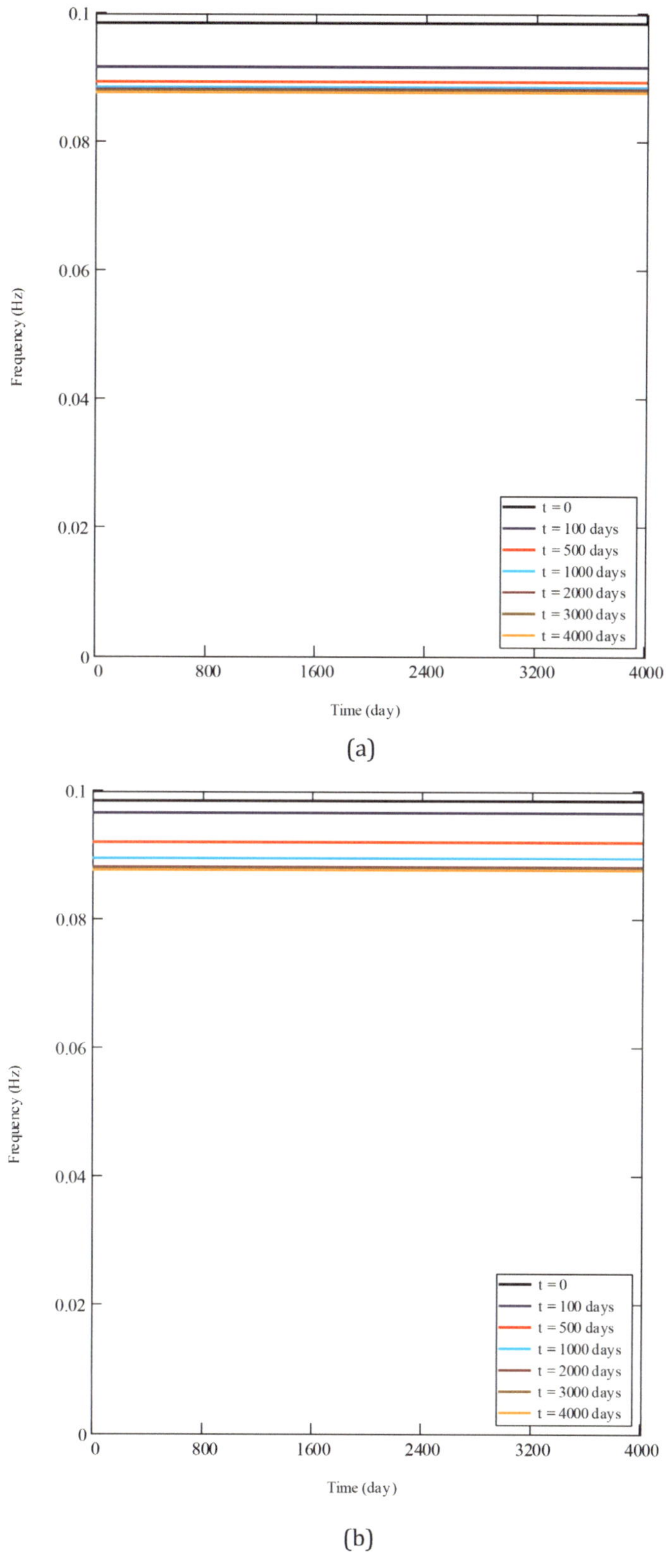

Figure 5.
Frequencies: (a) Eurocode model; (b) three-parameter model.

Time	Eurocode 2/three-parameter model (Hz)
0	0.098440
2000 days	0.087980
4000 days	0.087665

Table 1.
Frequencies for both models at selected instants.

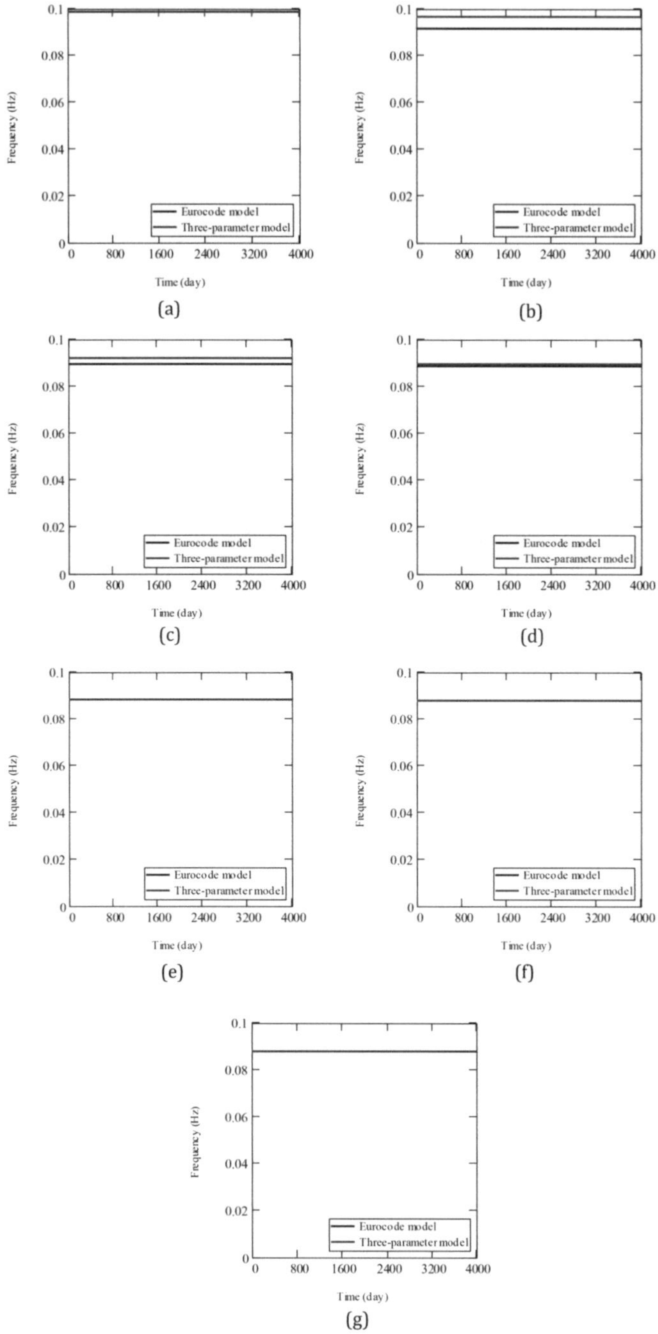

Figure 6.
Comparative of frequencies to different times. (a) t = 0. (b) t = 100 days. (c) t = 500 days. (d) t = 1000 days. (e) t = 2000 days. (f) t = 3000 days. (g) t = 4000 days.

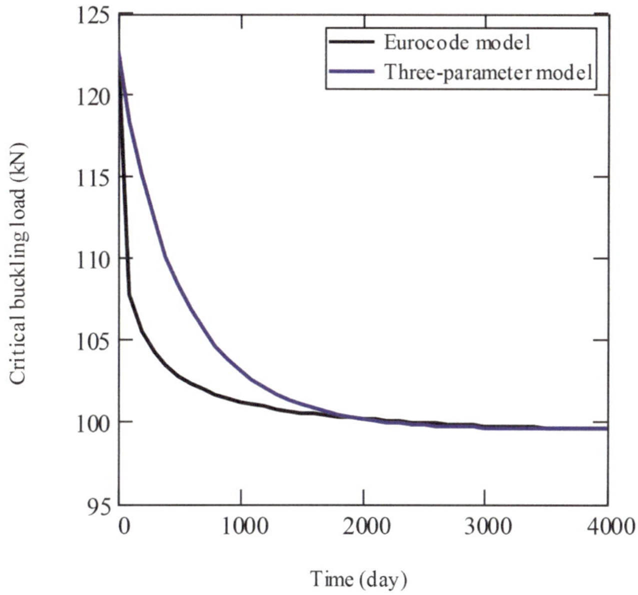

Figure 7.
Variation of the critical buckling load to different times.

6. Conclusions

- Because of the viscoelastic behavior of the material, the modulus of elasticity presents a variation along the time, reflecting on the structural frequency and critical buckling load.
- A three-parameter viscoelastic model has been adjusted to fit the same results as predicted by Eurocode creep criteria for a specific interval of time. The use of two parameters makes the adjustment process more flexible.
- It is important to stay clear that that adjustment does not lie on the adjustment of the modulus of the elasticity which does not have the same precision when observed for both models; for that a dynamic analysis is important.
- This article demonstrated the possibility of adjustment of a simple model to the standard one and how easy it is used for practical applications to calculate the first natural frequency as the critical buckling load.
- For future works, a programing routine for obtaining a finer adjustment of the curve between the viscoelastic rheological model of three parameters and that of the model for creep as predicted by Eurocode must be developed.
- Further, comparative analyses considering other values of environmental humidity should be also performed.

Author details

Alexandre de Macêdo Wahrhaftig
Department of Construction and Structures, Polytechnic School, Federal University of Bahia (UFBA), Salvador, BA, Brazil

*Address all correspondence to: alixa@ufba.br

References

[1] Awrejcewicz J, Krysko AV, Zagniboroda NA, Dobriyan VV, Krysko VA. On the general theory of chaotic dynamics of flexible curvilinear Euler-Bernoulli beams. Nonlinear Dynamics. 2015;**79**(1):11-29. DOI: 10.1007/s11071-014-1641-5

[2] Rayleigh L. Theory of Sound. New York: Dover Publications; 1877

[3] Bert CW. Application of a version of the Rayleigh technique to problems of bars, beams, columns, membranes, and plates. Journal of Sound and Vibration. 1987;**119**(2):317-326. DOI: 10.1016/0022-460X(87)90457-3

[4] Bert CW. Improved technique for estimating buckling loads. Journal of Engineering Mechanics. 1984;**110**(12):1655-1665. DOI: 10.1061/(asce)0733-9399(1984)110:12(1655)

[5] Gander MJ, Wanner G. From Euler, Ritz, and Galerkin to modern computing. Society for Industrial and Applied Mathematics. 2012;**54**(4):1-40. DOI: 10.1137/100804036

[6] Zuo H, Bi K, Hao H. Dynamic analyses of operating offshore wind turbines including soil structure interaction. Engineering Structures. 2018;**157**:42-62. DOI: 10.1016/j.engstruct.2017.12.001

[7] Awrejcewicz J, Krysko VA, Pavlov SP, Zhigalov MV, Kalutsky LA, Krysko AV. Thermoelastic vibrations of a Timoshenko microbeam based on the modified couple stress theory. Nonlinear Dynamics. 2019. In Press. DOI: 10.1007/s11071-019-04976-w

[8] Silva MA, Brasil RMLRF. Nonlinear dynamic analysis based on experimental data of RC telecommunication towers subject to wind loading. Mathematical Problems in Engineering. 2006;**2006**:1-10. DOI: 10.1155/MPE/2006/46815

[9] European Standard. EN 1992-1-1—Eurocode 2: Design of concrete structures—Part 1: General rules and rules for buildings; 2004

[10] de Macêdo Wahrhaftig A. Analysis of the first modal shape using two case studies. International Journal of Computational Methods. 2019;**16**(6):1840019-1/1840019-14. DOI: 10.1142/S0219876218400194

[11] de Macêdo Wahrhaftig A, Silva MA, Brasil RMLRF. Analytical determination of the vibration frequencies and buckling loads of slender reinforced concrete towers. Latin American Journal of Solids and Structures. 2019;**16**(5):1-31. DOI: 10.1590/1679-78255374

[12] Findley WN, Lai JS, Onaran K. Creep and Relaxation of Nonlinear Viscoelastic Materials, with an Introduction to Linear Viscoelasticity. New York: Dover Publications, Inc.; 1989

[13] Keramat A, Shirazi KH. Finite element based dynamic analysis of viscoelastic solids using the approximation of Volterra integrals. Finite Elements in Analysis and Design. 2014;**86**:89-100. DOI: 10.1016/j.finel.2014.03.010

[14] Mukudai J. Evaluation of linear and non-linear viscoelastic bending loads of wood as a function of prescribed deflections. Wood Science and Technology. 1983;**17**(3):203-216. DOI: 10.1007/BF00372319

[15] Kränkel T, Lowke D, Gehlen C. Prediction of the creep behaviour of bonded anchors until failure—A rheological approach. Construction and Building Materials. 2015;**75**:458-464. DOI: 10.1016/j.conbuildmat.2014.11.048

[16] Sellier A, Multon S, Buffo-Lacarrière L, Vidal T, Bourbon X, Camps G. Concrete creep modelling for structural applications: Non-linearity, multi-axiality, hydration, temperature and drying effects. Cement and Concrete Research. 2016;**79**:301-315. DOI: 10.1016/j.cemconres.2015.10.001

[17] Han B, Jiao YY, Xie HB, Zhu L. Creep of compression fly ash concrete-filled steel tubular members. Thin-Walled Structures. 2017;**114**:116-121. DOI: 10.1016/j.tws.2017.01.034

[18] Pascon JP. Large deformation analysis of functionally graded visco-hyperelastic materials. Computers & Structures. 2018;**206**:90-108. DOI: 10.1016/j.compstruc.2018.06.001

[19] Milašinović DD, Landovic A. Rheological-dynamical analogy for analysis of vibrations and low cycle fatigue in internally damped inelastic frame structures. Computers & Structures. 2018;**196**:76-93. DOI: 10.1016/j.compstruc.2017.11.001

[20] de Macêdo Wahrhaftig A, César SF, Brasil RMLRF. Creep in the fundamental frequency and stability of a slender wooden column of composite section. Revista Árvore. 2016;**40**(6): 1119-1130. DOI: 10.1590/ 0100-67622016000600018

[21] de Macêdo Wahrhaftig A, Brasil RMLRF, Nascimento LSMSC. Analytical and mathematical analysis of the vibration of structural systems considering geometric stiffness and viscoelasticity. Numerical Simulations in Engineering and Science. 2018;**1**:349-369. DOI: 10.5772/intechopen.75615

[22] Araújo JM. Comparative study of the simplified methods of Eurocode 2 for second order analysis of slender reinforced concrete columns. Journal of Building Engineering. 2017;**14**:55-60. DOI: 10.1016/j.jobe.2017.10.003

[23] Marin MC, El Debs MK. Contribution to assessing the stiffness reduction of structural elements in the global stability analysis of precast concrete multi-storey buildings. Revista IBRACON de Estruturas e Materiais. 2012;**5**(3):316-342. DOI: 10.1590/ S1983-41952012000300005

[24] Madureira EL, Siqueira TM, Rodrigues EC. Creep strains on reinforced concrete columns. Revista IBRACON de Estruturas e Materiais. 2013;**6**(4):537-560. DOI: 10.1590/ S1983-41952013000400003

[25] Kataoka LT, Bittencourt TN. Numerical and experimental analysis of time-dependent load transfer in reinforced concrete columns. Revista IBRACON de Estruturas e Materiais. 2014;**7**(5):747-760. DOI: 10.1590/ S1983-41952014000500003

Chapter 9

Dynamic Modelling by Bond Graph Approach of Convective Drying Phenomena

Hatem Oueslati, Salah Ben Mabrouk and Abdelkader Mami

Abstract

Drying operations play an important role in food industries. They are often the last operation of the process of manufacturing a product, with a strong influence on the final quality. The processes are numerous and depend on the type and amount of product to be dried and water to be evaporated, the desired final quality, or the desired functionality for the dried product. In this chapter, we present a modeling study of heat transfer during drying a moist agricultural product placed in a hot air flow in a tunnel dryer with partial solar heating. The bond graph approach has been used for system modeling, and it is an object-oriented graphical approach based on an energetic description between subsystems. Some drying tests have been carried out on tomatoes and the experimental results are compared with the theoretical results for the validation of the developed model.

Keywords: tunnel dryer, bond graph modelling, simulation, thermal transfer, temperature, moisture content

1. Introduction

Dehydration and drying involves the partial or complete elimination of the water contained in the food. Due to low water activity (a_w), microorganisms cannot proliferate, and most chemical or enzymatic deterioration reactions are slowed down. Solar drying, which is most commonly practiced in developing countries, can significantly improve the quality of the dried product compared to traditional drying methods such as sun and shade drying [1–3] while minimizing losses during the harvest period.

The shape or physical state of the product to be dried can help the industrialist to choose the right way or technique to practice drying. This requires a high heat input to cause the water to be easily removed from the wet product.

The drying device is a partial solar heating tunnel dryer (**Figures 1** and **2**); it uses solar energy as a source of renewable energy that represents an available market, furthermore to be practically free. Heat and mass transfers were investigated with the development of a mathematical model of the thin layer solar drying of food products, in a convective tunnel dryer.

Multiple numerical and experimental studies on convective drying were realized [4, 5]. On the one hand, they focus on understanding the phenomena that govern the internal migration of moisture, heat and mass transfer at the level of the

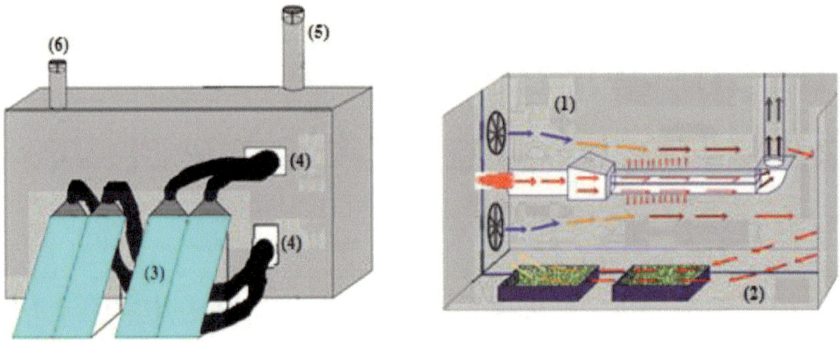

Figure 1.
Schematic view of drying device. (a) External view and (b) outside view. (1) Auxiliary heating system (two fans, burner, heat exchanger, air heating room), (2) drying chamber containing the trays, (3) solar module (four solar air collectors), (4) solar airflow vanes, (5) first chimney (smoke out), (6) second chimney (out moist air).

Figure 2.
Photo of the convective drying process.

air-product contact [6, 7]. On the other hand, they studied the physical analysis of the dryers and the optimization of their behaviour [8–10].

We recall that the drying is a complex phenomenon, and it includes many other phenomena that emerge from fluid mechanics, thermodynamics and heat and mass transfer [6, 11]. These phenomena are in turn playing a leading role in the drying process. The modelling tool used in this work is the bond graph approach [12]. It can model multi-disciplinary systems [13] whose their comportment is nonlinear. This is explained by the diversity of physical phenomena (thermal, mechanical, electrical, thermodynamic, chemical, etc.). The bond graph approach is a modelling tool that provides both the behaviour and the necessary analysis of models. It is a causal and modular energy modelling approach allowing the generation of ordinary differential equations. The bond graph approach can be used as a solution for the design and understanding of physical phenomena in complex industrial systems.

In process engineering, the use of the pseudo-bond graph approach is often widespread for reasons that have been widely justified [13–15], but for this approach, the product of effort and flow has not the dimension of power.

The first part of this work focuses on the development of a graphical model and the deduction of mathematical equations that describes the phenomena of heat and mass transfer for convective drying process. In the second part, we show an analysis of the results with the influence of different aerothermal parameters.

2. Drying kinetics

The theory of drying is described by Lewis [16], especially during thin film drying.

The most important parameter used in process drying is the drying rate equation [17]:

$$-\frac{dX}{dt} = k(X - Xe) \tag{1}$$

$$k = 0.00719 \exp\left(-\frac{130.64}{T_{ma}}\right) \tag{2}$$

T_{ch}, X and Xe are respectively the moist air temperature, the instantaneous moisture content and the equilibrium moisture content of the vegetable or the wet agricultural product. According to GAB model [18, 19] Xe is expressed by:

$$Xe = \frac{W_m C K a_w}{(1 - K a_w)[1 + (C - 1) K a_w]} \tag{3}$$

W_m, C and K are parameters related with air temperature by the following expressions:

$$W_m = 0.0014254 \exp\left(\frac{1193.2}{T_k}\right) \tag{4}$$

$$C = 0.5923841 \exp\left(\frac{1072.5}{T_k}\right) \tag{5}$$

$$K = 1.00779919 \exp\left(-\frac{43.146}{T_k}\right) \tag{6}$$

T_k and a_w are the air absolute temperature and the water activity.

3. Experimental device

The device studied is an indirect convective solar dryer composed of a tunnel and four solar collectors. The tunnel is of dimensions: 4 m × 2 m × 1.8 m, consisting of a drying chamber and an auxiliary heating system. The width of the drying chamber is 0.9 m with a chimney on the roof used for the evacuation of humid air. The chamber is made of galvanized sheet metal, the side walls of which are insulated with a layer of polyurethane 0.15 m thick (**Figures 1** and **2**).

The drying chamber contains perforated iron trays. The solar module is formed by four air collectors placed with an inclination between 36.7° and 45° for optimal operation. The ambient air was sucked in by the fans and warmed up then mixed by the solar air forming the drying air. The latter comes into contact with dry agricultural products placed in the trays (**Figures 1** and **2**).

The auxiliary heating system consists of a gas burner and an exchanger that transfers the heat produced by the burner into the incoming air. The tunnel dryer has two fans in the centre that blow hot air into the drying chamber.

The agricultural products placed on a tray are brought into contact with a flow of hot air; the moist air accumulates in the drying chamber and then escapes through the chimney due to the difference in vapour pressure between the room and the exterior for a type of non-forced flow.

For the study of this drying process, we take into consideration the following hypotheses:

- Dry air is often a mixture of solar air and auxiliary air for a hybrid drying operation (**Figure 3**).
- The contact surfaces between air and product remain constant during drying.
- The agricultural product to be dried is estimated as a thin layer of water.
- The phenomenon of heat transfer by conduction between the grains of products is neglected.
- The phenomena of condensation and radiation transfer are neglected inside and outside the drying chamber.
- The products to be dried are characterized by their surface temperature.

The phenomena that will be studied:

- Convective heat transfer and evaporation between air and product.
- Heat transfer by convection between the air and the inner walls of the chamber.
- Heat exchange by conduction between the internal air and the external environment.
- Evacuation of humid air to the outside through the chimney.

4. Modelling technique

The dynamic behaviour of the thermal processes is generally described by the nonlinear differential equations. Their formulation and resolution by the classic numerical methods are limited [20]. These equations are really associated with the physical phenomena such as storage and energy dissipation. The bond graph approach allows by their graphical description to show these energy exchanges in the system.

The bond graph modelling approach is a unified and causal approach applied to all types of dynamical systems; it allows the modellers to obtain the mathematical model in the form of a state equation easier than the classical modelling methods.

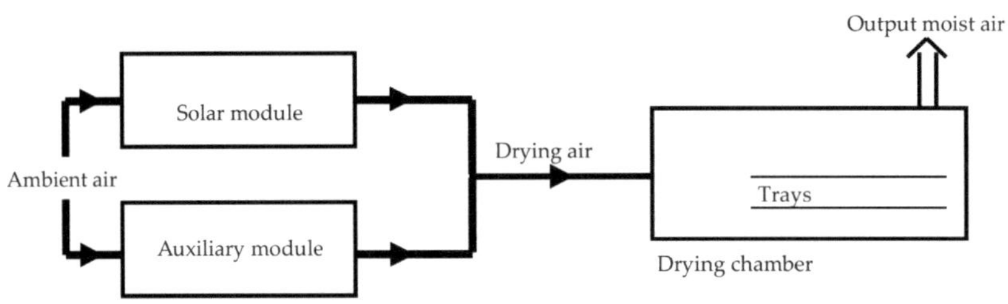

Figure 3.
Flow diagram of drying process in a tunnel dryer.

Moreover, to provide information on the structural properties of the studied system.

The first modelling step is to divide the global system into subsystems that exchange power with each other; this is the word-pseudo bond graph (**Figure 4**). The effort and flow variables are marked at the input and output of each subsystem. Depending on the physical phenomena that occur during drying and using the properties of the bond graph approach, the words are replaced by their corresponding elements, which lead to the complete models shown in **Figure 5**.

The pseudo-bond graph model of the studied tunnel dryer was developed based on the equations derived from the energy balances of the system.

The variables of effort and flux are respectively temperature (T) and heat flux ($\dot{Q}$), moreover, the pseudo bond graph model of **Figure 4** is constructed by five elements which are **Se**, **R**, **C**, **0** and **1**.

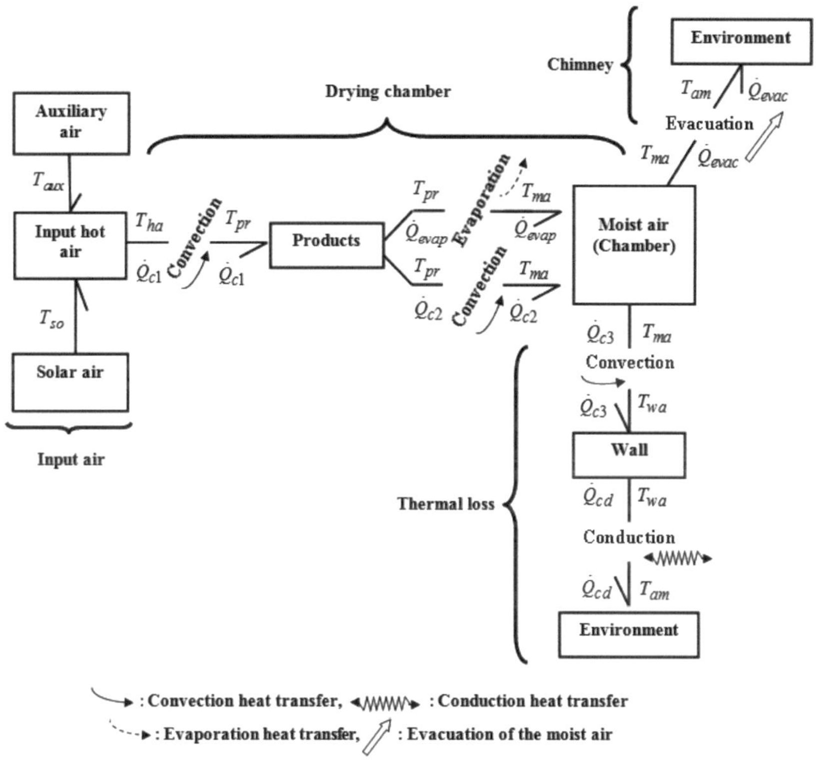

Figure 4.
Word pseudo bond graph model of the tunnel dryer.

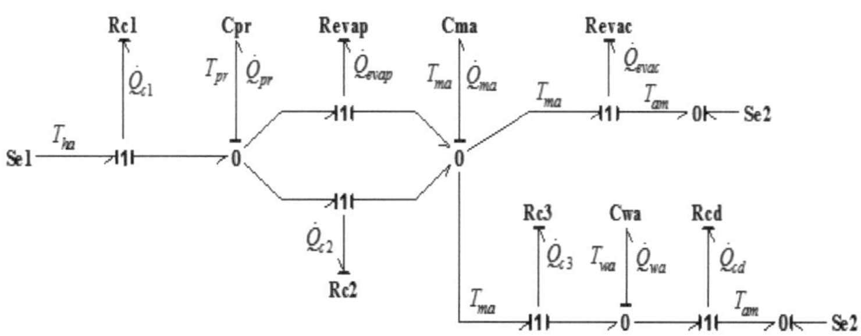

Figure 5.
Pseudo bond graph model of the tunnel dryer.

4.1 Effort sources

The two sources of effort used in this pseudo-bond graph model are **Se1** and **Se2** which respectively model the average temperature of the air drying T_{ha} and the ambient air temperature T_{am}.

$$T_{ha} = \frac{T_{aux} + T_{so}}{2} \tag{7}$$

4.2 C-fields

The energy storage phenomena are modelled by the **C** elements, the effort or flux variable is determined according to the attributed causality using the following relation:

$$e = \varphi_c^{-1} \int f dt \tag{8}$$

On the surface of the product modelled by the element **Cpr**, the corresponding temperature is expressed by:

$$T_{pr} = \frac{1}{C_{pr}} \int \dot{Q}_{pr} dt \tag{9}$$

$\dot{Q}_{pr}$ is the thermal heat flow accumulated on the surface of the product and C_{pr} is thermal capacity of products.

$$C_{pr} = m_{pr} C_{p,pr} \tag{10}$$

with m_{pr} being the mass and $C_{p,pr}$ being the specific heat of the product.

$\mathbf{C_{ma}}$ represents the accumulation of energy inside the drying chamber, the moist air temperature in the chamber is given by:

$$T_{ma} = \frac{1}{C_{ma}} \int \dot{Q}_{ma} dt \tag{11}$$

$\dot{Q}_{ma}$ is the thermal heat flow accumulated in the chamber and C_{ma} is the thermal capacity of the moist air.

$$C_{ma} = \rho_{ma} \, V_{ma} \, C_{p,ma} \tag{12}$$

ρ_{ma}, V_{ma} and $C_{p,ma}$ are respectively the density of the moist air, the volume of the chamber and the specific heat of the moist air.

The accumulation of energy on the inner wall of the chamber is modelled by the $\mathbf{C_{wa}}$ element; the temperature of the inner wall is expressed by:

$$T_{wa} = \frac{1}{C_{wa}} \int \dot{Q}_{wa} dt \tag{13}$$

$\dot{Q}_{wa}$ and C_{wa} are the thermal heat flow and the thermal capacity of the inner wall.

$$C_{wa} = \rho_{wa} V_{wa} C_{p,wa} \tag{14}$$

ρ_{wa}, V_{wa} and $C_{p,wa}$ are respectively the density, the volume and the specific heat of inner walls.

4.3 R-fields

The heat transfer phenomena in the thermal processes are modelled by the **R** elements and according to the causality attributed we determine the variable effort (e) and flux (f) by using the following relation:

$$f = \varphi_R^{-1}(e) \tag{15}$$

Note: the thermal resistance R is equal to the inverse of the heat transfer coefficient h ($R = 1/h$).

Convective heat transfer phenomena between hot air and agricultural product are modelled by $\mathbf{R_{c1}}$, where the convective heat flux is expressed by:

$$\dot{Q}_{c1} = \frac{1}{R_{c1}}(T_{ha} - T_{pr})A_{pr} = h_{c1}(T_{ha} - T_{pr})A_{pr} \tag{16}$$

A_{pr} and h_{c1} are respectively the area of product and the convective heat transfer coefficient.

$$h_{c1} = h_c = Nu\frac{\lambda_a}{D} \tag{17}$$

Nu is the Nusselt number determined by the Reynolds number (Re), which provides information on the flow regime.

λ_a and D are respectively the thermal conductivity of the air and the characteristic diameter of the layer of the product.

The Reynolds number is given by this relation:

$$\mathrm{Re} = \frac{U_a\, l}{v} \tag{18}$$

The airflow will certainly be turbulent in the dryer, to calculate the number of Nusselt we use the following correlation [21]:

$$Nu = 0,023.\,\mathrm{Re}^{0,8}.\mathrm{Pr}^{0,4}, \mathrm{Pr} = 0.7\ (\text{Prandtl number}). \tag{19}$$

$\mathbf{R_{c2}}$ models the convection heat transfer phenomena between the agricultural product and the moist air in the chamber, the convective heat flow is given by:

$$\dot{Q}_{c2} = \frac{1}{R_{c2}}(T_{pr} - T_{ma})A_{pr} = h_{c2}(T_{pr} - T_{ma})A_{pr} \tag{20}$$

$h_{c2} = h_c$ is the convective heat-transfer coefficient.

The phenomena of heat transfer by convection between the humid air and the internal wall of the chamber are modelled by the element $\mathbf{R_{c3}}$ in which the convective heat flux is expressed by:

$$\dot{Q}_{c3} = \frac{1}{R_{c3}}(T_{ma} - T_{wa})A_{wa} = h_{c3}(T_{ma} - T_{wa})A_{wa} \tag{21}$$

where A_{wa} is the area of the wall and $h_{c3} = h_c$ is the convective heat transfer coefficient between the moist air and the inner wall, taking as Nusselt number [22]:

$$Nu = 0,036.\,\mathrm{Re}^{4/5}.\mathrm{Pr}^{1/3} \tag{22}$$

$\mathbf{R_{cd}}$ models the conduction heat transfer phenomena the chamber walls and the external environment through the insulation, the conduction heat flow is given by:

$$\dot{Q}_{cd} = \frac{1}{R_{cd}}(T_{wa} - T_{ma})A_{wa} = h_{cd}(T_{wa} - T_{ma})A_{wa} \tag{23}$$

h_{cd} is the conductive heat-transfer coefficient across the insulation and estimated by:

$$h_{cd} = \frac{\lambda_i}{d_i} \tag{24}$$

λ_i is the thermal conductivity of the insulation and d_i is the average mean thickness of the insulation.

$\mathbf{R_{evap}}$ models the evaporation heat transfer phenomena from the agricultural product and the moist air in the drying chamber, the evaporation heat flow is given by:

$$\dot{Q}_{evap} = \frac{1}{R_{evap}}(T_{pr} - T_{ma})A_{pr} = h_{evap}(T_{pr} - T_{ma})A_{pr} \tag{25}$$

$$h_{evap} = 0.016 h_c \frac{[P(T_{pr}) - \gamma_{ma} P(T_{ma})]}{(T_{pr} - T_{ma})} \tag{26}$$

h_{evap} is the evaporative heat transfer coefficient [23] and γ is the relative decimal humidity and $P(T)$ the saturated vapour pressure given by Jain and Tiwari [22]:

$$P(T) = \exp\left[25.317 - \frac{5144}{T + 273.15}\right] \tag{27}$$

$\mathbf{R_{evac}}$ models the phenomenon of discharge of humid air to the outside through the chimney for a natural type of flow [24], the corresponding heat flow is given by:

$$\dot{Q}_{evac} = c_d A_e \sqrt{2g\Delta H}\Delta P \tag{28}$$

ΔP and ΔH are the difference in partial pressure and the difference in pressure head (m), respectively.

$$\Delta P = [P(T_{ma}) - \gamma_{am} P(T_{am})] \tag{29}$$

$$\Delta H = \frac{\Delta P}{\rho_a g} \tag{30}$$

4.4 (0.1)-junctions

Using the mathematical properties for the junctions (0.1):
$(\sum_i e_i = 0)$ for **1**-junctions.
$(\sum_i f_i = 0)$ for **0**-junctions.
The energy flow balances equations are:

- energy balance equation of the product

$$\dot{Q}_{pr} = \dot{Q}_{c1} - \dot{Q}_{c2} - \dot{Q}_{evap} \tag{31}$$

- energy balance equation of the moist air in the drying chamber

$$\dot{Q}_{ma} = \dot{Q}_{c2} + \dot{Q}_{evap} - \dot{Q}_{c3} - \dot{Q}_{evac} \tag{32}$$

- energy balance equation of the wall of the drying chamber

$$\dot{Q}_{wa} = \dot{Q}_{c3} - \dot{Q}_{cd} \tag{33}$$

The above equations determined by bond graph elements can be used to develop the detailed equations for the energy flow balance:

- energy balance equation of the product

$$C_{pr}\frac{dT_{pr}}{dt} = h_{c1}(T_{ha} - T_{pr})A_{pr} - (h_{c2} + h_{evap})(T_{pr} - T_{ma})A_{pr} \tag{34}$$

- energy balance equation of the moist air in the drying chamber

$$\begin{aligned} C_{ma}\frac{dT_{ma}}{dt} &= (h_{c2} + h_{evap})(T_{pr} - T_{ma})A_{pr} \\ &-h_{c3}(T_{ma} - T_{wa})A_{wa} - c_d A_e\sqrt{2g\Delta H}\Delta P \end{aligned} \tag{35}$$

- energy balance equation of the wall of the drying chamber

$$C_{wa}\frac{dT_{wa}}{dt} = h_{c3}(T_{ma} - T_{wa})A_{wa} - h_d(T_{wa} - T_{am})A_{wa} \tag{36}$$

5. Results and discussion

For the numerical evaluation of the thermal performance of the model developed for the tunnel dryer, the calculations were performed using the system parameters (**Table 1**). The [20-sim] software has been used for all simulations and is dedicated to bond graph simulation. Its use is quite simple and straightforward.

The measurements made were recorded in different operating variables following a series of experiments carried out in order to determine the performance of the studied tunnel dryer. This section presents the experimental results for several tomato drying operations and also a comparison with the theoretical results.

Parameters	Values	Parameters	Values
A_{pr}	2 (m^2)	γ_{ma}	0.65 (*dec*)
A_{wa}	3 (m^2)	γ_{am}	0.45 (*dec*)
A_e	0.00785 (m^2)	l	1 (m)
a_w	0.2	m_{pr}	10 (kg)
$C_{p,pr}$	4180 ($J/kg°C$)	X_{in}	16.1 kg kg^{-1} (*dry basis*)
$C_{p,ma}$	1006 ($J/kg°C$)	ρ_{ch}	1.16 (kg/m^3)
$C_{p,wa}$	860 ($J/kg°C$)	ρ_{wa}	2700 (kg/m^3)
c_d	0.6	λ_a	0.0262 ($W/m°C$)
D	0.15 (m)	λ_i	0.022 ($W/m°C$)
d_i	0.10 (m)	υ	2.10^{-5} (m^2/s)
g	9.8 (m/s^2)		

Table 1.
Values of parameters used in numerical simulation.

The interpretation of the results depends on the influence of the aerothermal parameters (velocity and temperature of drying air).

5.1 Influence of the hot air temperature

In a first step, we consider that the speed of drying air is constant and we only vary its temperature.

Figures 6 and 7 show the evolution of the product temperature and the temperature of the humid air, they reach after a certain time the temperature of the drying air. With the same conditions, we also represent the evolution of the moisture content of the product (**Figure 8**). Increasing the drying air temperature from 55 to 75°C is accompanied by a reduction in drying time. This is due to a potential increase to water evaporation.

Experimentally, for a temperature of 75°C, an air velocity of 2 m/s is sufficient 125 minutes to dry tomatoes. A decrease in temperature of 10° results in an increase

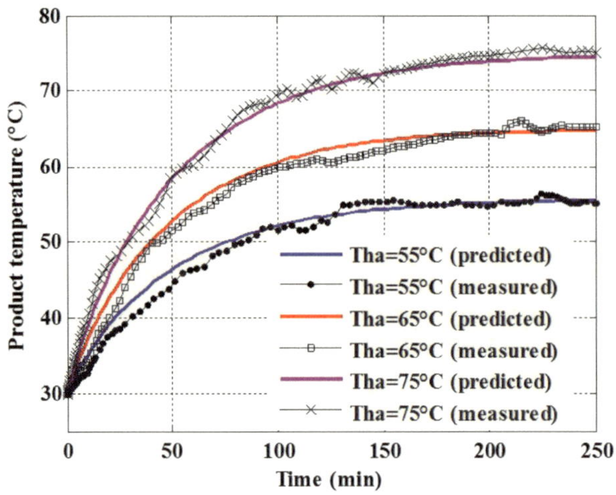

Figure 6.
Effect of the hot air temperature on the variation of the product temperature for U_a = 2 m/s *and* T_{am} = 30°C.

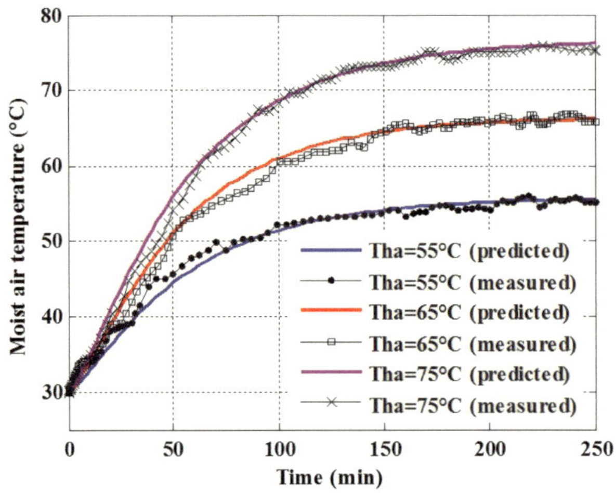

Figure 7.
Effect of the hot air temperature on the variation of the moist air temperature for U_a = 2 m/s *and* T_{am} = 30°C.

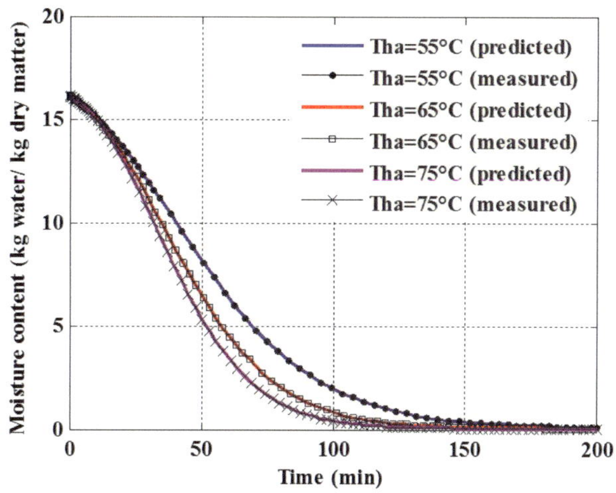

Figure 8.
Effect of the hot air temperature on the variation of the moisture content for U_a = 2 m/s and T_{am} = 30°C.

in drying time of 150 minutes to reach this content. Still with a lower temperature of 55°C, the drying time increases up to 180 minutes.

These results clearly show the influence of drying air temperature on the drying of agro-food products and are in good agreement with previous work [25, 26].

5.2 Influence of hot air velocity

For a constant drying air temperature 55°C and an increase in air velocity beyond 2 m/s does not show a good influence on the variation of the product temperatures and the humid air temperature in terms of reducing the drying time, this is clear in **Figures 9** and **10**.

The evolution of the moisture content is shown in **Figure 11** we see that the influence of the drying air velocity is less important than the drying air temperature because an increase in the air velocity has brought a small decrease in drying time [27, 28].

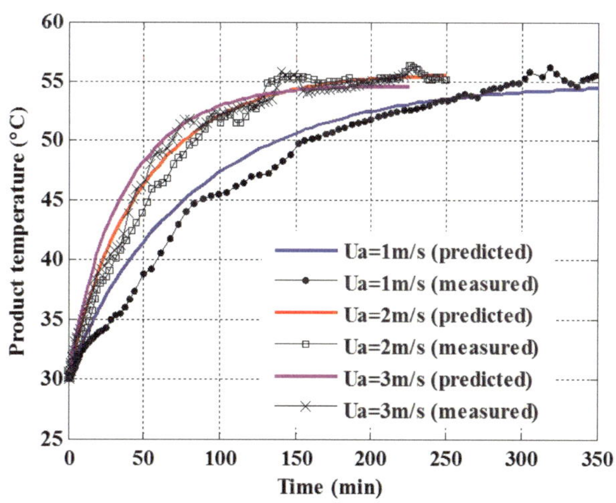

Figure 9.
Effect of the hot air velocity on the variation of the product temperature for T_{ha} = 55°C and T_{am} = 30°C.

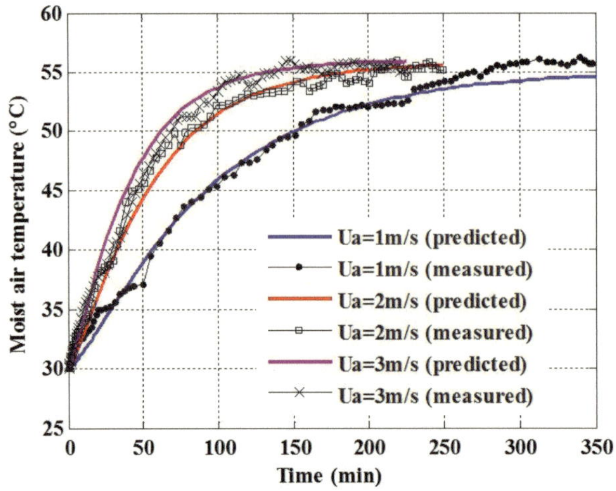

Figure 10.
Effect of the hot air velocity on the variation of the moist air temperature for T_{ha} = 55°C and T_{am} = 30°C.

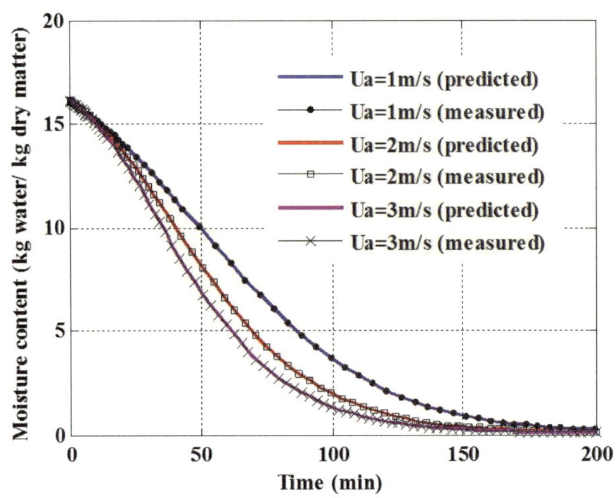

Figure 11.
Effect of hot air velocity on the variation of the moisture content for T_{ha} = 55°C and T_{am} = 30°C.

The predicted values of the different variables are in good agreement with the experimental values. The quality of the fit was determined using the Root Mean Square Errors (RMSE).

Variables	Influence of drying air temperature (Ua = 2 m/s)			Influence of drying air velocity (Tac = 55°C)		
	Tac = 55°C	Tac = 65°C	Tac = 75°C	Ua = 1 m/s	Ua = 2 m/s	Ua = 3 m/s
Product temperature	1.1249	1.1996	0.8625	1.3863	1.1249	1.2015
Internal air temperature	0.8034	0.9499	0.8665	0.8589	0.8034	0.9063
moisture content	$3.9665.10^{-4}$	$4.1778.10^{-4}$	$4.7045.10^{-4}$	$5.0070.10^{-4}$	$3.9665.10^{-4}$	$5.0572.10^{-4}$

Table 2.
RMSE values for different variables studied.

$$RMSE = \left(\frac{1}{N} \sum_{i=1}^{N} (X_{cal} - X_{mes})^2 \right)^{1/2} \quad (37)$$

We present in **Table 2** below the values of the mean squared error.

6. Conclusion

In this chapter, the bond-graph approach has been used for modelling a drying system with partially solar heating. This method provides reliable estimates of temperature distributions in the product and the moist air, also the moisture distributions in the product.

The geometry of the dryer, the physical properties of building materials, agricultural product and air are taken into account.

The influence of two aerothermal factors was studied to evaluate the performance of the dryer. The developed model can be adapted to other wet agricultural products as well as to other drying processes.

The challenge for the engineering designer is now to define optimal dryers, which provide a product of constant good quality. For this, the derived model of the tunnel dryer described by equations (1), (34), (35) and (36) will be used subsequently for the control of heat and mass transfer in drying process, which is important to enhance product quality such as colour and flavour.

Acknowledgements

This study was supported by the Research Program of Tunisian Ministry of High Education and Scientific Research.

Nomenclature

A	surface area (m^2)
a_w	water activity
C	thermal capacity (J/°C)
C_p	specific heat (J/kg°C)
c_d	discharge coefficient
D	characteristic diameter of the layer of the product (m)
d	mean thickness of the insulation (m)
g	gravitational acceleration (m/s^2)
h	heat-transfer coefficient (W/m^2°C)
k	drying constant (s^{-1})
l	characteristic length (m)
m	mass (kg)
P	saturated water vapour pressure (N/m^2)
$\dot{Q}$	heat flow plate (W)
T	temperature (°C)
T_k	air absolute temperature (K)
U	velocity (m/s)
V	volume (m^3)
X	product moisture content (kg/kg dry basis)

$\left(-\frac{dX}{dt}\right)$	drying rate (kg/kg dry basis. s)
Nu	Nusselt number
Pr	Prandtl number
Le	Lewis number

Greek letters

υ	kinematic viscosity of air (m^2/s)
ρ	density (kg/m^3)
λ	thermal conductivity (W/m°C)
γ	decimal relative humidity

Subscripts

a	air
aux	auxiliary air
am	ambient
c	convection
cd	conduction
e	equilibrium/exit
i	insulation
v	vapour
ma	moist air
$evap$	evaporation
evac	evacuation
pr	product
so	solar air
ha	hot air
wa	wall

Author details

Hatem Oueslati[1*], Salah Ben Mabrouk[1] and Abdelkader Mami[2]

1 Thermal Processes Laboratory, Research and Technology Centre of Energy (CRTEn), Hammam-Lif, Tunisia

2 Energetic Efficiency and Renewable Energies Application Laboratory (LAPER), Faculty of Sciences of Tunis, University of Tunis El Manar, Tunis, Tunisia

*Address all correspondence to: houeslati@gmail.com

References

[1] Pallav P, Atul K, Tara CK. Solar drying vs open sun drying: A framework for financial evaluation. Solar Energy. 2006;**80**(12):1568-1579. DOI: 10.1016/j.solener.2005.12.009

[2] Tunde Akintunde TY. Mathematical modeling of sun and solar drying of chilli pepper. Renewable Energy. 2011; **36**(8):2139-2145. DOI: 10.1016/j.renene.2011.01.017

[3] Rathore NS, Panwar NL. Experimental studies on hemi cylindrical walk-in type solar tunnel dryer for grape drying. Applied Energy. 2010;**87**(8):2764-2767. DOI: 10.1016/j.apenergy.2010.03.014

[4] Ben Mabrouk S, Khiari B, Sassi M. Modelling of heat and mass transfer in a tunnel dryer. Applied Thermal Engineering. 2006;**26**(17–18):2110-2118. DOI: 10.1016/j.fbp.2012.02.001

[5] Ben Mabrouk S, Benali E, Oueslati H. Experimental study and numerical modelling of drying characteristics of apple slices. Food and Bioproducts Processing. 2012;**90**(4):719-728. DOI: 10.1016/j.fbp.2012.02.001

[6] Bennamoun L, Belhamri A. Mathematical description of heat and mass transfer during deep bed drying: Effect of product shrinkage on bed porosity. Applied Thermal Engineering. 2008;**28**(17&18):2236-2244. DOI: 10.1016/j.applthermaleng.2008.01.001

[7] Yaldiz O, Ertekin C, Uzun H. Mathematical modelling of thin layer solar drying of sultana grapes. Energy. 2001;**26**(5):457-465. DOI: 10.1016/S0360-5442(01)00018-4

[8] Ben Mabrouk S, Belghith A. Development of the solar crop dryers in Tunisia. World Renewable Energy Congress VI. 2000:2206-2212. DOI: 10.1016/b978-008043865-8/50476-1

[9] Simate IN. Optimization of mixed-mode and indirect-mode natural convection solar dryers. Renewable Energy. 2003;**28**(3):435-453. DOI: S0960-1481(02)00041-1

[10] Oueslati H, Ben Mabrouk S, Marni A. Design and installation of a solar-gas tunnel dryer: Comparative experimental study of two scenarios of drying. In: Proceedings of the IEEE 2014 5th International Renewable Energy Congress (IREC); 25–27 March 2014; Hammamet, Tunisia. DOI:10.1109/IREC.2014.6826970

[11] Van Brakel J. Mass transfer in convective drying. In: Mujumdar AS, editor. Advances in Drying. Vol. Vol. 1. New York: Hemisphere Publication Corp; 1980

[12] Dauphin-Tanguy G. Les Bond Graphs. Paris: Hermes Science Publications; 2000

[13] Karnopp D, Margolis D, Rosenberg R. System Dynamics: A Unified Approach. New York: John Wiley & Son Inc; 1990

[14] Ould Bouamama B. Bond graph approach as analysis tool in thermofluid model library conception. Journal of the Franklin Institute. 2003;**340**(1):1-23. DOI: 10.1016/S0016-0032(02)00051-0

[15] Pichardo C, Delgado M. Pseudo bond graph model and simulation of an industrial flash separator. Simulation Modelling Practice and Theory. 2003; **11**(2):125-150. DOI: 10.1016/S1569-190X(03)00011-X

[16] Lewis WK. The rate of drying of solid materials. Journal of Industrial and Engineering Chemistry. 1921;**13**(5): 427-432. DOI: 10.1021/ie50137a021

[17] Lopez A, Iguaz A, Esnoz A, Vırseda P. Thin layer drying behaviour

of vegetable wastes from wholesale market. Drying Technology. 2000;**18** (4&5):985-994. DOI: 10.1080/07373930008917749

[18] Van Der Berg C. Description of water activity of foods for engineering purposes by means of the G.a.B model of sorption. In: McKenna M, editor. Engineering and Foods. Vol. 1. New York: Elsevier Applied Science Publishers; 1984. pp. 311-321

[19] Lopez A, Iguaz A, Esnoz A, Vırseda P. Modelling of sorption isotherms of dried vegetable wastes from wholesale market. Drying Technology. 2000;**18**(4&5):995-1006. DOI: 10.1080/07373930008917748

[20] Hadim H, Vafai K. Overview of current computational studies of heat transfer in porous media and their applications: Forced convection and multiphase heat transfer. In: Minkowycz WJ, Sparrow EM, editors. Advances in Numerical Heat Transfer. Vol. 2. Washington, DC: Taylor & Francis; 1999. pp. 291-329

[21] Pallet D, Fournier M, Themelin A. Modélisation, identification et simulation d'un séchoir solaire à bois. Revue de Physique Appliquée. 1987; **22**(1):1399-1409

[22] Jain D, Tiwari GN. Thermal aspects of open sun drying of various crops. Energy. 2003;**28**(1):37-54. DOI: 10.1016/S0360-5442(02)00084-1

[23] Goyal RK, Tiwari GN. Parametric study of a reverse flat plate absorber cabinet dryer: A new concept. Solar Energy. 1997;**60**(1):41-48. DOI: SOO3! 3-092X(96)00144-Z

[24] Jain D, Tiwari GN. Effect of greenhouse on crop drying under natural and forced convection II. Thermal modelling and experimental validation. Energy Conversion and Management. 2004;**45**(17): 2777-2793. DOI: 10.1016/j.enconman. 2003.12.011

[25] Laguerre JC, Lebert A, Trystram G, Bimbenet JJ. A compartmental model to describe drying curves of foodstuffs under variable conditions. In: Mujumdar AS, Filkova I, editors. Drying 91. Amsterdam, New York: Elsevier; 1991. pp. 361-368

[26] Azzouz S, Guizani A, Jomaa W, Belghith A. Moisture diffusivity and drying kinetic equation of convective drying of grapes. Journal of Food Engineering. 2002;**55**(4):323-330. DOI: S0260-8774(02)00109-7

[27] Bennamoun L, Belhamri A. Numerical simulation of drying under variable external conditions: Application to solar drying of seedless grapes. Journal of Food Engineering. 2006; **76**(2):179-187. DOI: 10.1016/j. jfoodeng.2005.05.005

[28] Toğrul IT, Pehlivan D. Modelling of thin layer drying kinetics of some fruits under open-air sun drying process. Journal of Food Engineering. 2004; **65**(3):413-425. DOI: 10.1016/j. jfoodeng.2004.02.001